Growth Curve Analysis and Visualization Using R

Daniel Mirman

Moss Rehabilitation Research Institute
Elkins Park, Pennsylvania

Drexel University,
Philadelphia, Pennsylvania

CRC Press
Taylor & Francis Group
Boca Raton   London   New York

CRC Press is an imprint of the
Taylor & Francis Group, an **informa** business

A CHAPMAN & HALL BOOK

Chapman & Hall/CRC
The R Series

Series Editors

John M. Chambers
Department of Statistics
Stanford University
Stanford, California, USA

Torsten Hothorn
Division of Biostatistics
University of Zurich
Switzerland

Duncan Temple Lang
Department of Statistics
University of California, Davis
Davis, California, USA

Hadley Wickham
Department of Statistics
Rice University
Houston, Texas, USA

Aims and Scope

This book series reflects the recent rapid growth in the development and application of R, the programming language and software environment for statistical computing and graphics. R is now widely used in academic research, education, and industry. It is constantly growing, with new versions of the core software released regularly and more than 5,000 packages available. It is difficult for the documentation to keep pace with the expansion of the software, and this vital book series provides a forum for the publication of books covering many aspects of the development and application of R.

The scope of the series is wide, covering three main threads:
- Applications of R to specific disciplines such as biology, epidemiology, genetics, engineering, finance, and the social sciences.
- Using R for the study of topics of statistical methodology, such as linear and mixed modeling, time series, Bayesian methods, and missing data.
- The development of R, including programming, building packages, and graphics.

The books will appeal to programmers and developers of R software, as well as applied statisticians and data analysts in many fields. The books will feature detailed worked examples and R code fully integrated into the text, ensuring their usefulness to researchers, practitioners and students.

Published Titles

Event History Analysis with R, *Göran Broström*

Statistical Computing in C++ and R, *Randall L. Eubank and Ana Kupresanin*

Reproducible Research with R and RStudio, *Christopher Gandrud*

Displaying Time Series, Spatial, and Space-Time Data with R,
Oscar Perpiñán Lamigueiro

Programming Graphical User Interfaces with R, *Michael F. Lawrence
and John Verzani*

Analyzing Baseball Data with R, *Max Marchi and Jim Albert*

Growth Curve Analysis and Visualization Using R, *Daniel Mirman*

R Graphics, Second Edition, *Paul Murrell*

**Customer and Business Analytics: Applied Data Mining for Business Decision
Making Using R**, *Daniel S. Putler and Robert E. Krider*

Implementing Reproducible Research, *Victoria Stodden, Friedrich Leisch,
and Roger D. Peng*

Dynamic Documents with R and knitr, *Yihui Xie*

CRC Press
Taylor & Francis Group
6000 Broken Sound Parkway NW, Suite 300
Boca Raton, FL 33487-2742

© 2014 by Taylor & Francis Group, LLC
CRC Press is an imprint of Taylor & Francis Group, an Informa business

No claim to original U.S. Government works

Printed on acid-free paper
Version Date: 20140108

International Standard Book Number-13: 978-1-4665-8432-7 (Hardback)

Library of Congress Cataloging-in-Publication Data

Mirman, Daniel, author.
   Growth curve analysis and visualization using R / Daniel Mirman.
      pages cm. -- (Chapman & Hall/CRC the R series)
   Includes bibliographical references and index.
   ISBN 978-1-4665-8432-7 (hardback)
   1. Biometry. 2. Regression analysis. 3. R (Computer program language) 4. Psychometrics. I. Title.

QH324.2.M57 2014
570.1'5195--dc23   2013048809

Visit the Taylor & Francis Web site at
http://www.taylorandfrancis.com

and the CRC Press Web site at
http://www.crcpress.com

Contents

List of Figures

List of Tables

Preface

About This Book

This book is intended to be a practical, easy-to-understand guide to carrying out growth curve analysis (multilevel regression) of time course or longitudinal data in the behavioral sciences, particularly cognitive science, cognitive neuroscience, and psychology. Multilevel regression is becoming a more and more prominent statistical tool in the behavioral sciences and it is especially useful for time course data, so many researchers know they *should* use it, but they do not know *how* to use it. In addition, analysis of individual differences (developmental, neuropsychological, etc.) is an important subject of behavioral science research but many researchers don't know how to implement analysis methods that would help them quantify individual differences. Multilevel regression provides a statistical framework for quantifying and analyzing individual differences in the context of a model of the overall group effects. There are several excellent, detailed textbooks on multilevel regression, but I believe that many behavioral scientists have neither the time nor the inclination to work through those texts. If you are one of these scientists – if you have time course data and want to use growth curve analysis, but don't know how – then this book is for you. I have tried to avoid statistical theory and technical jargon in favor of focusing on the concrete issue of applying growth curve analysis to behavioral science data and individual differences.

This book begins with a simple definition of time course or longitudinal data and a discussion of problems with analyzing separate time bins or windows using t-tests or ANOVAs. The first chapter will also provide a brief introduction to using `ggplot` to visualize time course data and describe how data need to be formatted for growth curve analysis and plotting. Chapter 2 will provide a basic overview of the structure of growth curve analysis (GCA) and how it addresses the challenges described in Chapter 1. This chapter will include the first two concrete examples of GCA, including the core analysis syntax using `lmer` and how to plot model fits. Chapter 3 will describe how to deal with change over time that is not linear, focusing on the potentially difficult challenge of selecting a growth curve model form. Chapter 4 will cover how to structure random effects, including how to analyze within-participant designs and whether participants should be treated as fixed or random effects. Chapter 5 will discuss how GCA (and regression more generally) uses categorical predictors and how to conduct multiple simultaneous comparisons among different levels of a factor. Chapter 6 will explain why binary outcomes require logistic models, how to implement logistic and quasi-logistic GCA, and the relative advantages and disadvantages of these approaches. Chapter 7 will discuss how to use GCA to analyze individual differences when there is a separate measure of those differences and how to extract estimates of individual effect sizes from the model itself.

Throughout this book, R code will be provided to demonstrate how to implement the analyses and to generate the graphs. Each chapter also ends with a few exercises so you can test your understanding. Chapter 8 presents the code for all of the key examples along with sample write-ups demonstrating how to report GCA results. The example datasets, code for solutions to the self-test exercises, and other supplemental code and examples can be found on the book website: http://www.danmirman.org/gca.

R

This book assumes minimal familiarity with R and no expertise in computer programming. If you are unfamiliar with R, then consider this a great time to start learning it. R (http://www.r-project.org/) is a free, open-source, cross-platform system for statistical computing and graphics. For beginners, *R in a Nutshell* by Joseph Adler is an accessible and comprehensive guide and Code School has developed an excellent interactive online tutorial called *Try R* (http://tryr.codeschool.com/). For those who are familiar with SAS or SPSS, Robert Muenchen's *R for SAS and SPSS Users* can help smooth the transition.

In addition to base R, there are many add-on packages that extend or simplify its functionality. For the purposes of this book, only two additional packages are required:

- lme4: This package implements linear mixed-effects (multilevel) regression and will be the primary tool for growth curve analysis.

- ggplot2: This package provides powerful and elegant graphing tools, which will be used throughout the book to plot data and model fits. The package website (http://ggplot2.org/) has documentation and other useful resources.

A few other packages are not required, but will prove useful, including plyr and reshape2 for manipulating data, stringr for working with character strings (text), multcomp for making multiple simultaneous comparisons among factor levels, and psych for miscellaneous tools and datasets that may be relevant to researchers in the psychological sciences. I also highly recommend using RStudio (http://www.rstudio.com/), which is a full-featured, cross-platform integrated development environment for R. RStudio has many features that make R a lot more user-friendly, including syntax highlighting, executing code directly from the source editor, workspace and data viewers, plot history with easy image export, and a package management and installation interface. For more advanced users, it also offers project management tools and one-click Sweave execution with PDF preview.

In addition to these relatively static resources, R has a very active and diverse online community, covering a huge range of topics (including the behavioral and social sciences) for all levels, from beginners to advanced developers. Cookbook for R (http://www.cookbook-r.com/) is a cookbook-style how-to wiki that covers a wide range of topics, including a detailed guide to plotting with `ggplot2`. R-bloggers (http://www.r-bloggers.com/) is a blog aggregator that pulls together R-related blog posts from hundreds of bloggers and is a great way to learn helpful tips and tricks and keep up with the cutting edge in the world of R. If you get stuck, Stackoverflow (http://stackoverflow.com/) is a question and answer site for professional and enthusiast programmers. Many thousands of R questions have already been asked and answered on the site, so the answer to your question might already be there. If it is not, then you should join the community and ask it – the answers are usually fast and helpful. For researchers in the language sciences, the R-lang mailing list (https://mailman.ucsd.edu/mailman/listinfo/ling-r-lang-l) can be a helpful resource and multilevel modeling is a frequent topic of discussion on that list.

Multilevel Regression

This book is meant to be a practical guide to implementing growth curve analysis, not a comprehensive textbook on multilevel regression or hierarchical linear modeling. For those interested in a deeper and broader statistical discussion of multilevel modeling, there are several excellent textbooks. Three that I recommend are

- Gelman, A., & Hill, J. (2007). *Data Analysis Using Multilevel/Hierarchical Models.* Cambridge University Press.

- Singer, J. D., & Willett, J. B. (2003). *Applied Longitudinal Analysis: Modeling Change and Event Occurrence.* Oxford University Press.

- Raudenbush, S. W., & Bryk, A. S. (2002). *Hierarchical Linear Models: Applications and Data Analysis Methods.* Sage Publications.

Acknowledgments

I am deeply grateful to Jim Magnuson, who set me on the path to using growth curve analysis, and J. Dixon, who was my first guide on that path. This book grew out of a series of tutorial workshops, so I also owe thanks to the people who invited me to teach those workshops: John Trueswell at the University of Pennsylvania, Sid Horton and Matt Goldrick at Northwestern University, and Rick Dale at the University of California - Merced. Before there were tutorials, there were individuals who asked to learn GCA and kept asking the right questions until I figured out the answers, particularly Solène Kalénine, Elika Bergelson, Chia-Lin (Charlene) Lee, and Allison Britt. Joe Fruehwald taught me to use the `ggplot2`, `plyr`, and `reshape` packages, which have proven invaluable. Finally, thanks to my wife, Jessica, who made this book, and my life, much better than it otherwise would have been.

1

Time course data

CONTENTS

1.1   Chapter overview

This chapter will describe the main problems that growth curve analysis is meant to address. First, it will define a particular kind of data, called *time course data* or *longitudinal data*, which involve systematic relationships between observations at different time points. These relationships pose problems for simple traditional analysis methods like t-tests.

Section 1.3 will discuss four kinds of problems and illustrate them with concrete examples. First, using separate analyses for individual time bins or time windows creates a trade-off between power (more data in each bin) and temporal resolution (smaller time bins). Second, flexibility in selection of time bins or windows for analysis introduces experimenter bias. Third, statistical thresholding ($p < 0.05$ is significant but $p > 0.05$ is not) makes gradual change look abrupt and creates the illusion that continuous processes are discrete. Fourth, there is no clear way to quantify individual differences, which are an important source of constraints for theories in the behavioral sciences.

Section 1.4 will provide a brief introduction to `ggplot2`, a powerful and flexible package for graphing data in R. Section 1.5 will distinguish between *wide* and *long* data formats and describe how to use the `melt` function to convert data from the wide to the long format, which is the right format for growth curve analysis and for plotting with `ggplot2`. The rest of this book will describe *growth curve analysis*, a multilevel regression method that addresses the challenges discussed in this chapter, provide a guide to applying growth curve analysis to time course data, and demonstrate how to use `ggplot2` to visualize time course data and growth curve model fits.

1.2 What are "time course data"?

Time course data are the result of making repeated observations or measurements at multiple time points. These sorts of data are also called *longitudinal* or, more generally, *repeated measures* data. Imagine that you measured a child's height annually from birth to 18 years old. You would have a series of 19 data points that describe how that child's height changed over time during those 18 years. In other words, the growth (height) *time course* for that child.

Two key properties distinguish time course data from other kinds of data. The first is that groups of observations all come from one source, which is called *nested data*. In the height example, the source was a particular child. If you repeated this procedure for another child, you would now have two nested series of data points corresponding to the two children in your study. The heights of two randomly selected children may be uncorrelated, but the height of a child at time t is strongly correlated with that child's height at time $t - 1$. Nested observations are not independent and this non-independence needs to be taken into account during data analysis. Capturing this nested structure allows quantifying the particular pattern of correlation among data points for an individual, which can reveal potentially interesting individual differences – a taller child compared to a shorter child, whether the child had an earlier or later growth spurt, etc.

In this example, the data were nested or grouped at the individual participant level. The grouping can also be at a higher level. For example, if you measured the weights of newborns at different hospitals every month for a year, you would have data grouped by hospital, rather than by individual child (each child was only weighed once, but each hospital's newborns were weighed every month). Groupings can also be at multiple levels; for example, if you followed those children as they grew, you would have measurements grouped by child and children grouped by hospital.

The second key property of longitudinal data is that the repeated measurements are related by a continuous variable. Usually that variable is *time*, as in the child growth example, but it can be any continuous variable. For example, if you asked participants to name letters printed in different sizes, you could examine the outcome (letter recognition accuracy) as a function of the continuous predictor *size*. On the other hand, if you had presented letters from different alphabets (Latin, Cyrillic, Hebrew, etc.), that would be a categorical predictor. For categorical predictors, one can only assess whether the outcome was different between different categories (for example, if recognition of Latin letters was better or worse than recognition of Cyrillic letters). For continuous predictors, one can do that kind of simple comparison, but it is also possible to assess the *shape* of the change – whether the relationship between letter recognition accuracy and letter size follows a straight line, or accuracy improves rapidly for smaller sizes and then reaches a plateau, or follows a U-shape. Because time is so frequently that critical continuous variable, this book will typically refer to these sorts of data as "time course data" even though the same issues apply to other continuous predictors.

As we will see, growth curve analysis (GCA) is a way to analyze nested data that takes the grouping into account and provides a way to quantify and assess the shapes of time course curves. Before getting into GCA, it will help to understand the challenges of analyzing time course data in a little more detail. That is, to understand why traditional methods like *t*-test and analysis of variance (ANOVA) are not well-suited to these sorts of data. To do that, the next section goes over some examples of the kinds of problems that come up when analyzing time course data.

1.3 Key challenges in analyzing time course data

How should time course data be analyzed? A simple approach is to apply traditional data analysis techniques like *t*-tests or ANOVAs. For example, we could independently compare conditions at each time bin or time window. This approach has a number of problems, which are easiest to demonstrate with concrete examples.

1.3.1 Trade-off between power and resolution

The data in Figure 1.1 are based on an experiment that examined whether words with high "transitional probability" (TP) would be learned faster than words with low TP (Mirman, Magnuson, Graf Estes, & Dixon, 2008). Word learning was predicted to be faster in the high TP condition than the low TP condition. The training trials were grouped into blocks to examine the gradual learning. The data in Figure 1.1 are the word "learning curves": the participants started out near chance (50% correct, because there are two response choices on each trial) and gradually got better, reaching about 90% correct at the end of 10 blocks of training trials. Importantly, it looks like this learning was faster for high TP words.

What kind of statistical test would provide the quantitative test of the effect of TP on word learning? Faster word learning means that participants in

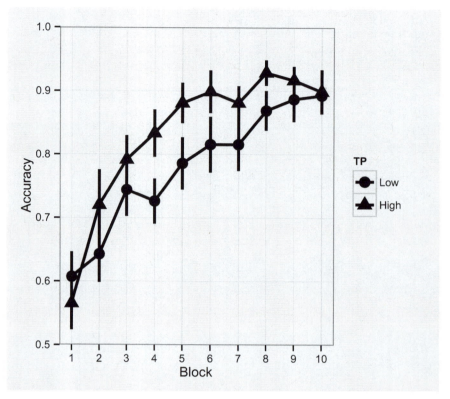

FIGURE 1.1
Effect of transitional probability (TP) on novel word learning.

the High TP condition generally have higher accuracy, so we could do a t-test comparing the High and Low TP conditions on overall accuracy. This turns out to be only marginally statistically significant ($t(54) = -1.69, p = 0.096$). We could do a repeated measures ANOVA with TP and Block as main effects and look for a TP-by-Block interaction, which would indicate that the TP effect differs across Blocks. For these data, we would get a strong main effect of Block (i.e., participants in both conditions learned the novel words: $F(9, 486) = 21.9$, $p < 0.001$) and a marginal main effect of TP (analogous to the overall t-test: $F(1, 54) = 2.87$, $p = 0.096$), but no hint of a TP-by-Block interaction ($F(9, 486) = 1.02$, $p = 0.42$).

Finally, we could run a series of t-tests comparing the TP conditions in each Block to see if any of those t-test comparisons are significant. Conducting such multiple comparisons increases the risk of a false positive result because the likelihood of observing $p < 0.05$ in any one of the 10 blocks is much higher than the nominal 5% false positive rate implied by $p < 0.05$. However, even without correcting for multiple comparisons, block-by-block t-tests do not seem to capture the difference in learning rate in a completely satisfying way (Table 1.1): there is just one block with a significant TP effect (block 4) and one block with a marginal TP effect (block 5). The problem is that

TABLE 1.1
Block-by-Block *t*-Test Results for Effect of TP on Novel Word Learning

Block	t	df	p
1	0.73	54.00	0.47
2	-1.09	54.00	0.28
3	-0.84	54.00	0.40
4	-2.08	54.00	0.04
5	-1.83	54.00	0.07
6	-1.51	54.00	0.14
7	-1.32	54.00	0.19
8	-1.60	54.00	0.11
9	-0.72	54.00	0.48
10	-0.13	54.00	0.89

gradual change over time can be difficult to detect: because it is gradual, the overall effect will be weak and because each time bin (trial block) has only a small amount of data, individual time bin comparisons are underpowered. In other words, there is a trade-off between statistical power, which requires more data and therefore larger time windows, and temporal resolution, which requires smaller time windows but undermines statistical power.

1.3.2 Possibility of experimenter bias

Consider the data in Figure 1.2, which are based on a spoken word comprehension experiment using eye-tracking (Magnuson, Dixon, Tanenhaus, & Aslin, 2007). The curves show the probability of fixating the named ("target") picture over time, starting at word onset. How could we verify that the crossover effect is statistically significant? We could divide the time range into an early time window (before the crossover point) and a later time window (after the crossover point) and test whether the condition effect is different in the two time windows. However, if that crossover was not predicted, or even if just the specific timing of the crossover point had not been predicted, then using the observed data themselves to define the analysis would constitute a case of "double-dipping," which increases the rate of false positives (e.g., Kriegeskorte, Simmons, Bellgowan, & Baker, 2009).

1.3.3 Statistical thresholding

A third problem with bin-by-bin analyses is that statistical thresholding can create spurious disagreements. That is, our inferential strategy of treating *p*-values less than 0.05 as fundamentally different from those greater than 0.05 (or 0.10, if the researcher is feeling generous) can turn (noisy) gradual changes

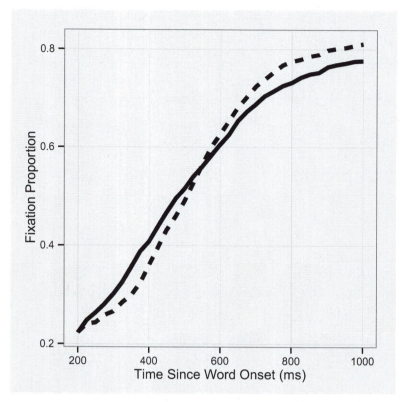

FIGURE 1.2
A crossover effect in target fixation probability.

into seemingly discrete, discontinuous differences. To illustrate this point, let's return to the word learning data, but to avoid any uncertainty, let's define the data in Figure 1.1 as the true underlying effect. Now we can generate simulated data with that shape and variability. In this scenario, we know for a fact that High TP words were learned faster, as shown in the top panel of Figure 1.3. Twenty simulated replications of the experiment — all using the same true underlying effect — were generated and analyzed with block-by-block t-tests to test the reliability of this approach.

The results of these block-by-block t-tests are shown in the bottom panel of Figure 1.3 with filled squares indicating a statistically significant advantage for the High TP condition ($p < 0.05$) without correcting for multiple comparisons. In 15 of the 20 replications the difference was significant in at least one block, but in 5 there was not a single training block that showed a statistically significant advantage for the High TP condition despite the fact that there was a true advantage. Although failures to replicate are to be expected to some degree (e.g., Francis, 2012), it is not hard to imagine skeptics taking this 25% rate of failure to replicate as evidence that there is no effect. Their skepticism

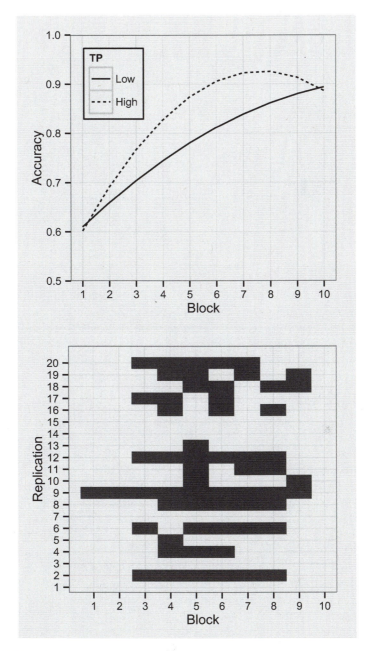

FIGURE 1.3
Top: Underlying data pattern for simulated replications of the effect of TP on novel word learning. Bottom: Results of block-by-block t-tests for the 20 simulated replications. Filled squares indicate statistically significant differences ($p < 0.05$, uncorrected).

could be further bolstered by the striking variability across replications. Based on the shape of the underlying model from which the data were generated, one would expect that the blocks 4-7 would show the differences most reliably. To some degree this was true — significant differences were detected most frequently in these blocks — but that frequency was only 50-60%. Another problem is that the number of blocks with $p < 0.05$ differed radically from replication to replication (0 - 9 blocks) and in about half of these replications, the $p < 0.05$ blocks were not contiguous (for example, in replication #10, the TP effect was significant in Blocks 5 and 9, but not in 6, 7, or 8). All of this inconsistency arose despite the data having been generated from a single, continuous underlying effect (top panel of Figure 1.3). Given concerns about replicability in the psychological sciences (e.g., the November 2012 issue of *Perspectives on Psychological Science*), we need statistical methods that are more robust than this.

1.3.4   Individual differences

In the previous example, each of the simulated replications had the same underlying true effect and just differed due to simulation of a different random set of individual participants. Traditional analyses like t-tests and ANOVA assume random variation among individual participants and stop there, limiting theories to describing a hypothetical prototypical individual. However, we can ask a deeper question: what is the source of this variability among individuals? This is an important question because individual differences provide unique constraints on our theories. Insofar as individuals differ from that prototype, this tells us something about how the system (cognitive, psychological, behavioral, neural, etc.) is organized. A good theory should not just account for the overall average behavior of a system, but also for the ways in which the system's behavior varies. For example, a good theory of human language processing should not only account for how typical college students process language, but also how language processing develops from infancy through adulthood into old age and how it breaks down, both in developmental and acquired disorders. All of this variability is not random — it is structured by the nature of the system — but we can't understand that structure unless we can quantify individual differences. Traditional data analysis methods like t-tests and ANOVAs do not provide a method for doing this.

   To sum up, time course data entail some unique data analysis challenges. (1) Bin-by-bin analyses force a trade-off between statistical power and temporal resolution. (2) Comparing different time windows has the potential to introduce experimenter bias. (3) Normal variability can produce spurious disagreements about time course effects due to statistical thresholding in bin-by-bin analyses. (4) Traditional methods like t-tests and ANOVAs do not provide a meaningful way to quantify individual differences, which are an important source of constraints on theories. Growth curve analysis (GCA) provides a way to address these challenges. Before statistically analyzing data it is important to be able to visually inspect it and to get it into the right format. The next two sections will introduce a set of tools for visualizing time course data in R and describe how to get data into the right format for analysis and plotting.

1.4 Visualizing time course data

So far, we've been talking about the challenges of analyzing time course data. Time course data can also pose challenges for visualization, so it is important to have powerful and easy-to-use tools for graphing. The R package `ggplot2` is particularly good for two reasons: (1) the "Grammar of Graphics" approach provides a flexible and powerful framework for visualizing data and (2) summary statistics like means and standard errors can be computed "on the fly." There are many excellent guides and tutorials for `ggplot2`, so this section will only provide a brief introduction, focusing on time course data.

The "Grammar of Graphics" approach is somewhat different from typical graphing frameworks. In programs like Excel or MATLAB®, you specify a set of (x, y) coordinate pairs and the style (color, shape, etc.) of symbols that will be placed at those locations. In `ggplot`, you first assign variables in your data to properties of the graph. These assignments or mappings are called the *aesthetics* of your graph. Then you select "geometries," or *geoms* — points, lines, bars, etc. — for those aesthetics. Once you get the hang of this approach, it provides a consistent, easy to manipulate, and intuitive framework for visualizing your data. Here are some examples using the built-in `Orange` data set, which contains growth data for 5 orange trees.

In this book, each line of R code will begin with a > character, as it would in the R console. You should be able to type or copy the code (without the line-start character) directly into R and run it. In order to use the `ggplot2` package, you first need to load it

```
> library(ggplot2)
```

The main function is `ggplot`, which takes two inputs (function inputs are usually called *arguments*). The first input is the data frame that you want to plot; in our case this is `Orange`. The second input uses the `aes` function to set up the aesthetic mappings between variables in the data and visual properties of the graph. By default, the first input to `aes` is mapped to the x-axis and the second input is mapped to the y-axis (you can also specify this explicitly). After that, you need to specify which aesthetic is being mapped. The code below will produce a simple scatterplot of the `Orange` data.

```
> ggplot(Orange, aes(age, circumference, shape=Tree)) +
    geom_point()
```

The first part

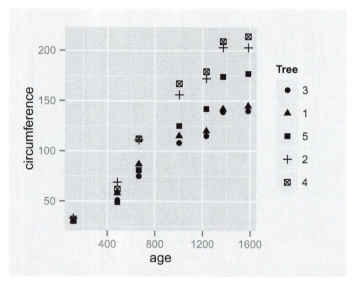

FIGURE 1.4
A simple scatterplot.

```
> ggplot(Orange, aes(age, circumference, shape=Tree))
```

sets up the **ggplot** object using the **Orange** data set and defining the aesthetics: **age** on the x-axis, **circumference** on the y-axis, and the different trees represented by different symbol shapes. **ggplot** will automatically recognize that **age** and **circumference** are continuous variables and **Tree** is a categorical variable and treat them appropriately. The second part

```
> + geom_point()
```

tells **ggplot** that these mappings should be realized using points, which makes the scatterplot in Figure 1.4. This use of the "+" operator in **ggplot** syntax may seem a little strange at first, since it is not exactly performing an addition operation, but it will quickly become second nature.

To add lines connecting the points (Figure 1.5), all you need to do is add the line geom using **geom_line()**. Note that even though **shape** doesn't apply to the lines, the grouping of points by **Tree** is inherited by the line geom:

```
> ggplot(Orange, aes(age, circumference, shape=Tree)) +
    geom_point() + geom_line()
```

One of the advantages of using **ggplot** is that it is easy to switch between different types of graphs. For example, switching to a line graph with trees represented by different line types is simply a matter of changing the mapping of the **Tree** variable to the **linetype** aesthetic (Figure 1.6).

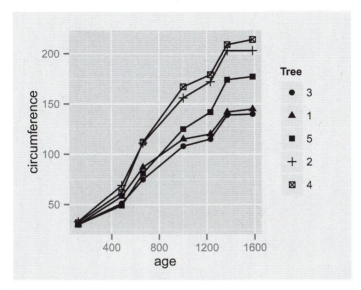

FIGURE 1.5
A line graph with points.

```
> ggplot(Orange, aes(age, circumference, linetype=Tree)) +
    geom_line()
```

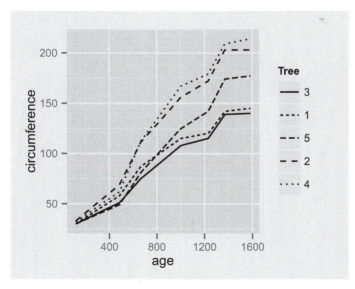

FIGURE 1.6
A line graph.

This advantage is particularly nice when you want to switch between black-and-white figures for publication and color figures for presentations. All you need to do is change the mapping of the `Tree` variable to the `color` aesthetic, which will be automatically applied to both the points and the lines (see book website for examples):

```
> ggplot(Orange, aes(age, circumference, color=Tree)) +
    geom_point() + geom_line()
```

Because `Tree` is a categorical variable, `ggplot` will pick contrasting colors for the trees. If `Tree` had been a continuous variable, `ggplot` would have used a continuous color gradient.

Sometimes data are too complicated for plotting on a single graph and you might want to create subplots or *small multiples*, which are called *facets* in `ggplot`. Facets are a series of similar plots that show different aspects of the data in a way that makes them easy to compare. In `ggplot`, facets are essentially another aesthetic dimension – you just need to specify which variables should be faceted, albeit with slightly different syntax. In Figure 1.7 each tree is plotted in a separate facet.

```
> ggplot(Orange, aes(age, circumference)) +
    facet_wrap(~ Tree) + geom_line()
```

By default, `facet_wrap` creates a ribbon of small plots, one for each of the unique values in the given variable. To create a grid of facets with one variable defining the rows and another defining the columns, use

```
> facet_grid(row_variable ~ column_variable)
```

Another great feature of `ggplot` is that it can compute summary statistics "on the fly." If you wanted to plot the mean growth pattern across all of the trees, in many graphing programs, you would first have to compute that mean and then plot those mean data. Within `ggplot` you can compute that mean using `stat_summary`, which is very convenient because you can easily look at individual data, then group data, then re-group or exclude individuals and check how this affects the overall patterns, etc. In other graphing packages this kind of exploration leaves you with a proliferation of different data sets and it can be hard to remember which one is which. With `ggplot`, you can do all this exploration using just your original data set. The way to do this is to use `stat_summary` and tell it which summary statistic to compute and which geom to use to visualize it. For example, this will plot the means of all trees as a line (Figure 1.8):

```
> ggplot(Orange, aes(age, circumference)) +
    stat_summary(fun.y=mean, geom="line")
```

To add an indication of the standard error, you need to compute that summary statistic and map it to an appropriate geom, such as `pointrange` or `errorbar` (Figure 1.9).

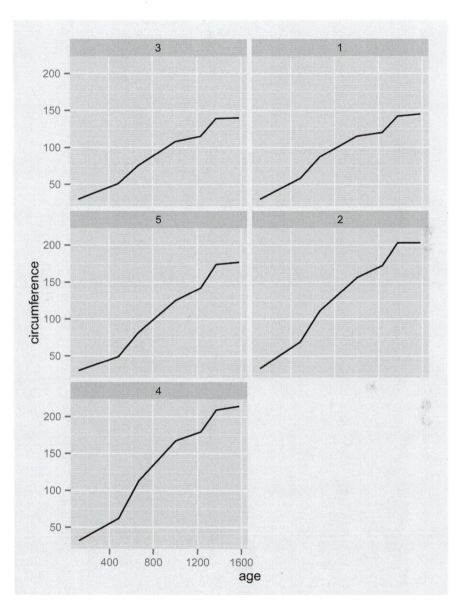

FIGURE 1.7
Facet plot.

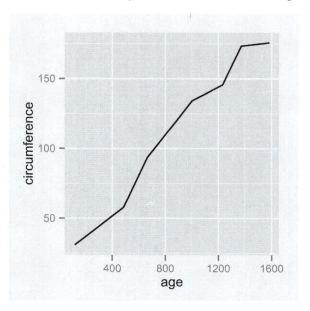

FIGURE 1.8
Summary plot: A line representing the mean at each age.

```
> ggplot(Orange, aes(age, circumference)) +
    stat_summary(fun.y=mean, geom="line") +
    stat_summary(fun.data=mean_se, geom="pointrange")
```

There are a variety of summary functions that come built-in with `ggplot` and you can write your own fairly easily. Note that the built-in standard error function computes basic between-subject standard errors. If you have within-subject variables and want to plot within-subject standard errors (e.g., Baguley, 2012), you will need to write an appropriate summary function.[1]

When creating figures for manuscripts or presentations, you may want to override various `ggplot` defaults to make the graphs look exactly how you want and to export the graphs at a particular size and resolution. The `ggsave` function is a convenient way to write `ggplot` graphs to a variety of image formats and allows specifying image dimensions and resolution. Here is an example of how to create a publication-ready graph of the `Orange` data and the result is plotted in Figure 1.10. There are, of course, many other customization options, some of which will be demonstrated in examples throughout this book.

```
> ggplot(Orange, aes(age, circumference)) +
    stat_summary(fun.y=mean, geom="line") +
    stat_summary(fun.data=mean_se, geom="pointrange",
```

---

[1]Examples can be found on the *Cookbook for R* website: http://www.cookbook-r.com/Graphs/Plotting_means_and_error_bars_(ggplot2)/

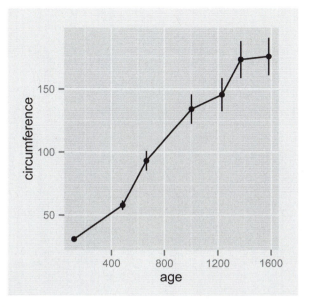

FIGURE 1.9
Summary plot: The line shows the mean; the vertical line through each point shows the standard error.

```
                  size=1) +
    theme_bw(base_size=10) +
    labs(x="Age (days since Dec. 31, 1968)",
         y="Trunk Circumference (mm)")
  > ggsave("Orange.pdf", width=3, height=3, dpi=300)
```

1.5   Formatting data for analysis and plotting

If you are new to R, you may want to start with an introductory book or tutorial to become familiar with the basic data types and data exploration functions (see Preface for recommendations). In this section, the focus will be on getting your data into the right format for the analysis and plotting methods described in the rest of the book. To begin, it will be helpful to distinguish between two ways of formatting nested data: *wide* and *long* data formats. In a wide format (also sometimes called *multivariate*), each row corresponds to a participant (or other individual observational unit) and each observation is in a separate column. For example, consider a subset of the **affect** data

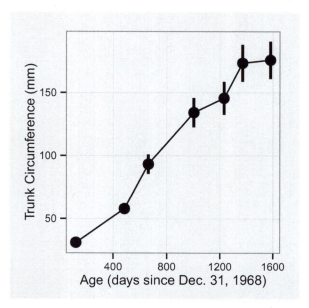

FIGURE 1.10
Average growth of orange trees. Error bars indicate ±SE.

set (from the `psych` package) that shows how negative affect is influenced by viewing one of four different 9-minute film excerpts:

- Sad: a *Frontline* documentary about liberation of concentration camps

- Threat: the 1978 horror film *Halloween*

- Neutral: *National Geographic* film about the Serengeti plain

- Happy: the 1989 comedy *Parenthood*

(for details see Rafaeli & Revelle, 2006, or the data set documentation using `?affect`)

```
> summary(affect.subset)
    Study           Film            NA1               NA2
  maps:160   Sad     :83    Min.    : 0.00    Min.    : 0.00
  flat:170   Threat :78    1st Qu.: 1.00    1st Qu.: 0.00
             Neutral:85    Median : 2.00    Median : 3.00
             Happy  :84    Mean    : 3.69    Mean    : 4.65
                           3rd Qu.: 6.00    3rd Qu.: 7.00
                           Max.    :28.00    Max.    :30.00
```

Participants were tested before and after the films, so each participant (row) has two observations of negative affect: pretest (`NA1`) and posttest (`NA2`). This is most clear if we examine the first few rows of the data frame using the `head` function:

```
> head(affect.subset)
  Study    Film NA1 NA2
1  maps Neutral   2   4
2  maps Neutral   4   5
3  maps Neutral   2   1
4  maps Neutral   0   2
5  maps Neutral  13  13
6  maps     Sad   1   2
```

This "wide" data format is convenient for some analyses, such as a paired-samples *t*-test, but for many analyses, including repeated measures ANOVA and growth curve analysis, and for plotting with `ggplot2`, the data need to be rearranged into a long (or *univariate*) format. In a long format each row corresponds to a single observation and the outcome data are all in one column (as opposed to two or more columns corresponding to measurement occasions). This essentially means stacking the values in the `NA1` and `NA2` columns (hence the term *long* format), creating a new variable that will identify whether the value came from `NA1` or `NA2`, and repeating the appropriate values from the other columns (`Study` and `Film`). This can be done manually, but the function `melt` from the `reshape2` package provides a powerful and easy-to-use interface for doing this conversion. The syntax of the `melt` function makes the distinction between two kinds of variables:

- `id` variables: information that identifies the observation, such as Subject, Time, Condition, etc. These are separate columns in the original data and will remain separate columns.

- `measure` variables: measurement values, such as reaction times, negative affect scores, etc. These are separate columns in the original data and will be converted into a new id variable consisting of the column names and a value column consisting of the values.

The `melt` function takes a data frame (it also works on arrays and lists, but only the data frame version is relevant for our purposes), a list of `id` variables, and a list of `measure` variables. If only the `id` or only the `measure` variables are specified, all other columns will be assumed to correspond to the other category. Here is how to use `melt` to convert the `affect.subset` data frame from wide to long format specifying that `Study` and `Film` are the `id` variables and `NA1` and `NA2` are the `measure` variables:

```
> affect.melt <- melt(affect.subset, id=c("Study","Film"),
                       measure=c("NA1", "NA2"))
> summary(affect.melt)
```

```
   Study            Film        variable        value
maps:320    Sad     :166    NA1:330    Min.    : 0.00
flat:340    Threat  :156    NA2:330    1st Qu.: 0.00
            Neutral:170                Median : 2.00
            Happy   :168               Mean    : 4.17
                                       3rd Qu.: 6.00
                                       Max.    :30.00
```

The `summary` shows that there is now a new column called `variable` that contains the column names (`NA1` and `NA2`) and a column called `value` that contains the negative affect scores. These new column names are the default values, but they can be specified in the call to `melt` to produce more informative variable names. Also, since the original data contain only id and measure variables, we only need to specify one set (in this example, the id variables) and can let the other variables become the other set by default:

```
> affect.melt <- melt(affect.subset, id=c("Study", "Film"),
                    variable.name="Test",
                    value.name="Negative.Affect")
> summary(affect.melt)
   Study            Film        Test       Negative.Affect
maps:320    Sad     :166    NA1:330    Min.    : 0.00
flat:340    Threat  :156    NA2:330    1st Qu.: 0.00
            Neutral:170                Median : 2.00
            Happy   :168               Mean    : 4.17
                                       3rd Qu.: 6.00
                                       Max.    :30.00
```

Now we can use `ggplot` to plot the data (Figure 1.11):

```
> ggplot(affect.melt, aes(Film, Negative.Affect, shape=Test))+
    stat_summary(fun.data=mean_se, geom="pointrange") +
    scale_shape_manual(values=c(1,16)) +
    theme_bw(base_size=10)
```

1.5.1   A note on data aggregation

Throughout this book we will consider data that have just one observation at the lowest level of nesting. That is, just one observation per participant per time point (per condition, for within-participant manipulations). Raw data do not always have this kind of structure. In particular, there are often multiple trials per condition, which need to be either aggregated into a single observation (e.g., by averaging across trials) or modeled as an additional level of nesting. Similarly, the sampling frequency might be much faster than behavioral changes (for example, an eye-tracker might record eye position every 2ms, but planning and executing an eye movement typically takes about 200ms),

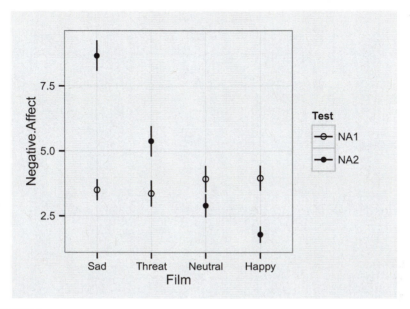

FIGURE 1.11
Effect of film type on negative affect. Error bars indicate ±SE.

which can produce many identical observations and lead to false positive results. Aggregating such oversampled data into larger time bins (e.g., 50ms) or otherwise downsampling the raw data may help to avoid this problem. The optimal approaches to aggregating or pre-processing data vary widely depending on the nature of the data, the research domain, and the research questions, so it is not possible to offer a simple strategy here. Just keep in mind that the analysis and visualization methods described in this book will assume that the relevant aggregation and pre-processing have already been done; if they have not, the results might be misleading or difficult to interpret.

1.6 Chapter recap

This chapter started with a simple definition of *time course data*, emphasizing two key properties: observations are nested within individuals and related by a continuous predictor. It then discussed problems with analyzing separate time bins or windows using *t*-tests or ANOVAs: (1) defining time bins creates a trade-off between statistical power and temporal resolution and (2) introduces experimenter bias; (3) normal variability in the data is exaggerated by statistical thresholding; and (4) the data can provide interesting insights into individual differences, but *t*-tests and ANOVAs do not provide a meaningful way to quantify individual differences. The final two sections provided a brief introduction to using `ggplot2` to visualize time course data and described how to use the `melt` to get data into the right format for analysis and plotting. The next chapter will provide a conceptual overview of growth curve analysis and begin demonstrating how to apply it to behavioral data.

1.7   Exercises

1. The `ChickWeight` data set contains data on the effect of diet on early growth of chicks (`?ChickWeight` for more details). Use this data set to practice plotting longitudinal data:

   (a) Make a scatterplot that shows weights of individual chicks as a function of time and diet.

   (b) Make a summary plot that shows average weight over time for each diet.

   (c) Add an indicator of standard error to the averages.

   (d) Customize the plot with more informative axis labels.

   (e) Make color and black-and-white versions of the plot.

   (f) Use `ggsave` to export the plot as an image file and specify an image size and resolution.

2. The `USArrests` data set contains violent crime arrests (per 100,000 residents) in each of the 50 states in the USA in 1973 and the percent of the population of each state that lived in urban areas (`?USArrests` for more details on this data set, and try `?state` for other information about US states).

   (a) Convert the `USArrests` data set from a wide to a long format so that instead of separate variables for each crime type (Murder, Assault, Rape), there is one variable that identifies the crime type and one variable that contains the rates for each crime type for each state.

   (b) Make a scatterplot showing the relationship between each type of violent crime rate and percent of population living in urban areas.

   (c) Plot the violent crime types in separate panels (tip: try using the `scales` and `nrow` or `ncol` options to customize the panels).

2

Conceptual overview of growth curve analysis

CONTENTS

2.1   Chapter overview

The previous chapter described the challenges of analyzing time course data. This chapter will provide a conceptual overview of *multilevel regression*, which is a way to address those challenges. First, regression methods explicitly model time as a continuous variable, which is the natural way to quantify change over time. In a regression framework, it is possible to move beyond modeling just linear changes over time – Chapter 3 will describe how to capture non-linear change over time. Second, multilevel regression provides a way to explictly model the nested structure of time course data. A core aspect of multilevel regression methods is that they simultaneously quantify both group-level and individual-level patterns within a single analysis framework. In other words, multilevel regression allows one to simultaneously describe the overall group pattern (as in traditional methods) and to describe how individual participants deviate from that pattern. The result is a quantitative description of the data that follows the nested structure of the data, including individual differences. Quantifying individual differences opens up entire new avenues for scientific investigation and provides new constraints on accounts of the data. Third, the regression framework provides an easy way to examine the effects of covariates, thus allowing more complex and informative analyses than simple *t*-tests.

    A few words about terminology: multilevel regression is a family of methods that is part of the broader field of longitudinal and repeated-measures data analysis techniques and goes by different names, including "multilevel regression," "hierarchical regression," or "hierarchical linear modeling" (HLM), and "growth curve analysis." Throughout this book "multilevel regression" will be used for general discussion of multilevel regression methods and "growth curve analysis" (or more simply "GCA") will be used when discussing specific applications to the kinds of data we encounter in the psychological and neural sciences.

This book is meant to be a practical guide to using multilevel regression in the psychological and neural sciences. Statistical theory and proofs will be side-stepped as much as possible, but we can't proceed without a basic overview of the underlying math. This chapter will provide that basic overview – just enough to understand how to apply these methods to the kinds of data we typically encounter, but without getting into all of the mathematical details. Gelman and Hill (2007) and Singer and Willett (2003) provide more complete technical treatments. After the basic overview, we will walk through two simple (linear) applications of growth curve analysis to behavioral data. These examples will include full R code for fitting the models, evaluating them, and plotting the data and model fits.

2.2   Structure of a growth curve model

To introduce the basic structure of a growth curve model, let's start with the simple linear case illustrated in Figure 2.1. We can mathematically describe the depicted relationship between outcome variable Y and Time as

$$Y = \beta_0 + \beta_1 \cdot Time \tag{2.1}$$

where β_0 is the intercept (i.e., value of Y when $Time = 0$) and β_1 is the slope (i.e., the average change in Y for every unit of Time). In the context of a regression model, that general relationship is elaborated to become a model of the individual observations Y_{ij} for individual i at Time j:

$$Y_{ij} = \beta_{0i} + \beta_{1i} \cdot Time_j + \varepsilon_{ij} \tag{2.2}$$

where ε_{ij} is the residual error, that is, the amount that actual observation Y_{ij} differs from the predicted value (as shown in Figure 2.1). Residual errors are generally assumed to be *independent and identically distributed*, meaning that all ε_{ij} come from the same distribution and that any particular ε_{ij} value does not provide any information about other values of ε_{ij}.

In a multilevel regression framework, Equation 2.2 describes the "Level 1" model. The reason it is called *multilevel* regression is that we can define a "Level 2" model of the Level 1 coefficients β_{0i} and β_{1i}. For example, imagine

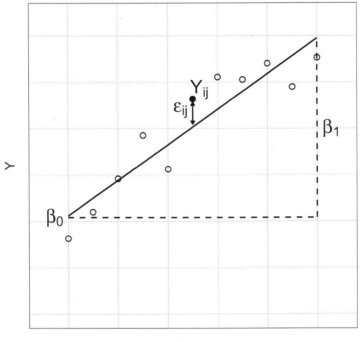

FIGURE 2.1
Linear regression schematic: The solid line is the linear regression line for the individual observations illustrated by the points. β_0 is the intercept; β_1 is the slope.

that we ran a study with two conditions: a control condition that will serve as the baseline and an experimental condition that we'll call condition C. Now the Level 2 model of the Level 1 intercept coefficient β_{0i} will be:

$$\beta_{0i} = \gamma_{00} + \gamma_{0C} \cdot C + \zeta_{0i} \tag{2.3}$$

where γ_{00} is the baseline value of β_{0i}, γ_{0C} is the *fixed effect* (also sometimes called *structural effect*) of condition C on the intercept, and ζ_{0i} is the random (also called *residual* or *stochastic*) deviation from that baseline for individual i. This Level 2 model does two things. First, it allows condition C to have a unique intercept that is different from the baseline intercept. Conceptually, this is roughly like a t-test comparing condition C to the baseline and, typically, these are the effects of primary interest. Second, it defines a structure for the random variation in the intercept: it says that all observations within the set indexed by i should have the same intercept, which can differ for other values of i. In other words, the random effect structure captures the nested structure of the data by specifying that all of the observations indexed by a particular value of i correspond to a single intercept, which is different from the intercepts for other values of i. In a typical case, i could refer to individual participants in a study and ζ_{0i} would vary randomly across participants.

It is important to distinguish the roles of fixed effects (γ) and random effects (ζ) in the model. One easy way to think about this is that fixed effects are *interesting in themselves*. In other words, they are reproducible, fixed properties of the world. In studies of psychological and neural processes, these are typically the experimental manipulations; for example, control vs. intervention, placebo vs. active drug(s), nouns vs. verbs, working memory load, or age of participants, and so on. Mathematically, the critical property is that fixed effect coefficients are estimated independently and are unconstrained. So, for example, each level of working memory load can have whatever γ parameter best fits the data without any consideration of the other levels.

In contrast, random effects (ζ) correspond to the randomly sampled observational units over which you intend to generalize. The two most common cases are that these will be either individual participants or individual items. For example, in a study comparing processing of nouns vs. verbs, the random effects could correspond to the particular nouns and verbs that were selected by the researcher with the intention of making general claims about all nouns and verbs. Conceptually, the notion is that these individual observational units are sampled randomly from some population. The standard way to implement this notion mathematically is by the constraint that random effects are drawn from a normal distribution with a mean of 0. Unlike fixed effects, which are unconstrained and independent, random effects are interdependent because they are meant to reflect random variation in the population. As mentioned in Chapter 1, this random variation may reflect interesting individual differences and Chapter 7 will discuss in detail how this information can be extracted and analyzed.

At the most general level, the goal of regression analysis is to find the parameters that best describe the data. To do this, we have to define what we mean by "best." One very powerful and flexible definition is to say that we want parameters that maximize the likelihood of observing the actual data, which is called *maximum likelihood estimation* or MLE. For standard (not multilevel) linear regression, the traditional *ordinary least squares* (OLS) regression algorithm can solve an equation to find the MLE parameter estimates, assuming the errors are normally distributed. For multilevel models this direct method is not possible (there is no closed-form solution), so an iterative algorithm is used, which tries to gradually converge to the MLE parameter estimates. However, it is not guaranteed to converge and the likelihood of convergence failure tends to increase with the complexity of the model, especially of the random effects structure.

Since the goal of MLE is to maximize the likelihood of observing the actual data, its goodness of fit is evaluated using the *log-likelihood* (LL) of the data given the estimated parameters. Other measures of model fit, such as

R^2, have an inherently meaningful interpretation (proportion of variance accounted for), but LL is only meaningful in the context of (meaningful) comparisons. That is, we can ask whether adding a critical parameter to the model improves the model fit (LL) by a significant amount. This comparison is called the *likelihood ratio test* (sometimes abbreviated LRT) because the difference of two log-values is equal to the log of the ratio of those values. The critical statistic is -2 times the change in log-likelihood ($-2 \cdot \Delta LL$), which is distributed as χ^2 with degrees of freedom equal to the number of parameters added to the model. Note that the LRT is used to evaluate the effect of adding or removing one or more parameters; that is, when one model contains a subset of the parameters of the other model, which are called *nested* models.

There exist other measures of model fit, such as the Akaike information criterion (AIC) and the Bayesian information criterion (BIC) both of which are computed from the log-likelihood with an adjustment for the number of free parameters. In principle, these measures can be used to compare non-nested models, but the difficulty is that the number of free parameters is not well-defined for multilevel models. Specifically, it is not clear whether each random effect estimate (i.e., each value of ζ_i) should be considered a free parameter or, because the random effects are constrained to come from a normal distribution with a mean of 0, only the estimated variance of that distribution is truly a free parameter. For these reasons, we will just use the LRT for model comparisons.

2.3 A simple growth curve analysis

2.3.1 Effect of amantadine on recovery from brain injury

We can now start implementing growth curve analysis using R. This first example will use an illustrative subset of data from a randomized placebo-controlled study of the effect of amantadine on recovery from brain injury (Giacino et al., 2012; thanks to Joseph Giacino, John Whyte, and their research team for sharing these data). The study tested patients who were in a vegetative or minimally conscious state 4 to 16 weeks after a traumatic brain injury. Following baseline assessment, patients were randomly assigned to receive either amantadine or placebo for 4 weeks (within-participants manipulations will be covered in Chapter 4). The primary outcome was the rate of functional recovery over the 4 weeks of treatment, as measured using the Disability Rating Scale (DRS), which ranges from 0 to 29, with higher scores indicating greater disability.

As a first step, it is useful to just inspect the data set:

```
> summary(amant.ex)
    Patient          Group          Week          DRS
 1008   :  5   Placebo  :85   Min.   :0   Min.   : 7.0
```

```
1009    :  5    Amantadine:65    1st Qu.:1    1st Qu.:17.0
1017    :  5                     Median :2    Median :20.5
1042    :  5                     Mean   :2    Mean   :19.3
1044    :  5                     3rd Qu.:3    3rd Qu.:22.0
1054    :  5                     Max.   :4    Max.   :28.0
(Other):120
```

The summary shows that there are four variables in the data:

- `Patient`: a patient ID code.

- `Group`: a group variable that has two levels, `Placebo` and `Amantadine`, indicating which drug the patient received.

- `Week`: the week on which each observation was made, which ranges from 0 to 4.

- `DRS`: the outcome variable disability rating score.

Note that the **summary** function gives distributional information for the continuous numeric variables `Week` and `DRS` (mean, median, and range) and number of observations for the categorical variables `Patient` and `Group`.

```
> ggplot(amant.ex, aes(Week, DRS, shape=Group)) +
    stat_summary(fun.data=mean_se, geom="pointrange")
```

Both groups seem to exhibit an approximately linear pattern of recovery (Figure 2.2), so we can try to model these data with a simple linear model. If the data did not look like straight lines, then straight-line models wouldn't describe the data properly. The next chapter describes how to handle more complex data shapes. Adapting the Level 1 model from Equation 2.2 to these data:

$$DRS_{ij} = \beta_{0i} + \beta_{1i} \cdot Week_j + \varepsilon_{ij} \tag{2.4}$$

We'll evaluate the effect of amantadine by including Group as a fixed effect in the Level 2 models:

$$\beta_{0i} = \gamma_{00} + \gamma_{0Group} \cdot Group + \zeta_{0i} \tag{2.5}$$

$$\beta_{1i} = \gamma_{10} + \gamma_{1Group} \cdot Group + \zeta_{1i} \tag{2.6}$$

The fixed effects γ_{0Group} and γ_{1Group} capture the systematic differences between groups (that is, the effect of the drug) in terms of starting DRS (γ_{0Group}) and rate of recovery (γ_{1Group}). The random effects ζ_{0i} and ζ_{1i} capture (random) individual variablity among patients in terms of their starting severity (ζ_{0i}) and rate of recovery (ζ_{1i}).

To implement the analysis, we will use the `lmer` function from the `lme4` package. First we need to load that package:

```
> library(lme4)
```

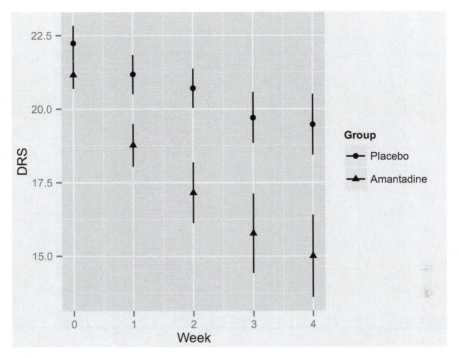

FIGURE 2.2

Recovery as measured by reduction in the Disability Rating Scale (DRS) score over the four weeks of the study for the placebo and amantadine groups. Vertical lines indicate ± SE.

We start out with a "base" model of recovery that has the Level 1 structure and the random effects, but no fixed effects of Group:

```
> m.base <- lmer(DRS ~ 1 + Week + (1 + Week | Patient),
                 data=amant.ex, REML=FALSE)
```

Let's unpack the `lmer` syntax. The first argument is the model formula. The tilde (~) operator can be read as "is a function of": the left side specifies the outcome variable and the right side specifies the predictors. In this case, `DRS` is the outcome. The first set of predictors are the fixed effects

```
> 1 + Week
```

which correspond to the intercept (indicated by the 1) and the slope (`Week`). The second set of predictors are the random effects

```
> (1 + Week | Patient)
```

which specify `Patient`-level random variablity in the baseline severity (intercept: 1) and rate of recovery (slope: `Week`). In other words, it is the model described in Equations 2.4, 2.5, and 2.6 but without γ_{0Group} and γ_{1Group}. Novice programmers may be unfamiliar with the vertical line "|" in the random effects specification – this is called the "pipe" and is typically located above the **Enter** key and produced by **Shift+**. The second argument tells **lmer** which data to use and the final argument tells it to use maximum likelihood estimation to fit the model (as opposed to *restricted* maximum likelihood estimation, which is not always appropriate for likelihood ratio tests).

To add a fixed effect of Group on the intercept we simply add **+ Group** to the fixed effects portion of the model formula

```
> m.0 <- lmer(DRS ~ 1 + Week + Group + (1 + Week | Patient),
             data=amant.ex, REML=FALSE)
```

and, similarly, we can add the effect of Group on the linear term by using the interaction syntax **+ Week:Group**

```
> m.1 <- lmer(DRS ~ 1 + Week + Group + Week:Group +
             (1 + Week | Patient), data=amant.ex, REML=FALSE)
```

Once we have each of these three models, we can test whether adding the Group fixed effects improved model fit using the **anova** function, which will do the model comparison for any number of models:

```
> anova(m.base, m.0, m.1)
Data: amant.ex
Models:
m.base: DRS ~ 1 + Week + (1 + Week | Patient)
m.0: DRS ~ 1 + Week + Group + (1 + Week | Patient)
m.1: DRS ~ 1 + Week + Group + Week:Group + (1 + Week | Patient)
        Df AIC BIC logLik deviance Chisq Chi Df Pr(>Chisq)
m.base   6 622 641   -305      610
m.0      7 623 644   -304      609  1.63      1      0.202
m.1      8 619 643   -302      603  5.56      1      0.018 *
---
Signif. codes:  0 '***' 0.001 '**' 0.01 '*' 0.05 '.' 0.1 ' ' 1
```

The **anova** output gives a reminder of the data set that was analyzed and the models that were compared and then the key model comparisons, including the χ^2 test for improvement in model fit. The model goodness (log-likelihood) is in the **logLik** column and the test statistic $(-2 \cdot \Delta LL)$ is in the **Chisq** column. The results indicate that adding a fixed effect of Group on the intercept did not improve model fit (**m.base - m.0** comparison: $\chi^2(1) = 1.63, p = 0.2$) and adding a fixed effect of Group on the slope did improve model fit (**m.0 - m.1** comparison: $\chi^2(1) = 5.56, p = 0.018$). In other words, the two groups (placebo vs. amantadine) did not differ significantly at baseline (intercept), but they did differ in their rate of recovery (slope).

The **summary** function will give a detailed summary of the model, including the actual parameter estimates and their standard errors:

```
> summary(m.1)
Linear mixed model fit by maximum likelihood ['lmerMod']
Formula: DRS ~ 1 + Week + Group + Week:Group +
                    (1 + Week | Patient)
   Data: amant.ex

     AIC      BIC    logLik deviance
  619.26   643.35   -301.63   603.26

Random effects:
 Groups    Name        Variance Std.Dev. Corr
 Patient   (Intercept) 3.198    1.788
           Week        0.698    0.836    0.45
 Residual              1.331    1.154
Number of obs: 150, groups: Patient, 30

Fixed effects:
                        Estimate Std. Error t value
(Intercept)              22.059     0.485     45.5
Week                     -0.700     0.221     -3.2
GroupAmantadine          -1.428     0.737     -1.9
Week:GroupAmantadine     -0.831     0.336     -2.5

Correlation of Fixed Effects:
            (Intr) Week   GrpAmn
Week         0.224
GroupAmntdn -0.658 -0.148
Wk:GrpAmntd -0.148 -0.658  0.224
```

By default, lmer treats the reference level of a factor as the baseline and estimates parameters for the other levels. In this case, that means that the (Intercept) parameter refers to the placebo group's intercept (22.06), the Week parameter refers to the placebo group's slope (-0.70), the GroupAmantadine parameter refers to the amantadine group's intercept relative to the placebo group's intercept (1.43 points lower), and the Week:GroupAmantadine parameter refers to the amantadine group's slope relative to the placebo group's slope (0.83 points per week faster decrease).[1]

You might notice that the parameter estimates provided by the summary do not have *p*-values. For these sorts of models the best test of statistical significance is nested model comparisons like the ones conducted using the anova function – where we test the improvement in model fit due to adding a single parameter. For complex models this can be impractical, so later chapters will discuss estimating parameter-specific *p*-values in more detail.

---

[1]These results are slightly different from those reported by Giacino et al. (2012) because this example only analyzed a simplified subset of their data.

2.3.2   Simplified model formula syntax

For expository purposes it was helpful to use a more verbose model formula
for `m.1`, but for convenience it can be shortened in two ways. First, `lmer`
will assume an intercept term by default, so it is not necessary to specify it
explicitly (0 can be used to override this default and eliminate the intercept
term, that is, force it to have a value of 0). Second, the asterisk operator (*)
can be used to specify "all main effects and interactions," so `Week + Group +`
`Week:Group` can be shortened to `Week*Group`. Together, this means that the
syntax for `m.1` can be substantially shortened:

```
> m.1_shorter <- lmer(DRS ~ Week*Group + (Week | Patient),
                      data=amant.ex, REML=FALSE)
```

and here is the summary to verify that the result is exactly the same:

```
> summary(m.1_shorter)

Linear mixed model fit by maximum likelihood ['lmerMod']
Formula: DRS ~ Week * Group + (Week | Patient)
   Data: amant.ex

     AIC      BIC   logLik deviance
  619.26   643.35  -301.63   603.26

Random effects:
 Groups    Name        Variance Std.Dev. Corr
 Patient   (Intercept) 3.198    1.788
           Week        0.698    0.836    0.45
 Residual              1.331    1.154
Number of obs: 150, groups: Patient, 30

Fixed effects:
                      Estimate Std. Error t value
(Intercept)            22.059     0.485     45.5
Week                   -0.700     0.221     -3.2
GroupAmantadine        -1.428     0.737     -1.9
Week:GroupAmantadine   -0.831     0.336     -2.5

Correlation of Fixed Effects:
            (Intr) Week   GrpAmn
Week         0.224
GroupAmntdn -0.658 -0.148
Wk:GrpAmntd -0.148 -0.658  0.224
```

2.3.3 Plotting model fit

It is always a good idea to plot the model fit with the observed data. The log-likelihood model fit statistic provides relative goodness of fit information (i.e., the effect of group on slope improved model fit), but this doesn't tell us how good the fit actually was. In particular, it doesn't tell us whether the statistical effect corresponds to a scientifically important difference in the data. It is very easy to add model fits to observed data using `ggplot`: we just need to specify a new mapping for the y-variable (model-fitted values instead of observed values; the other mappings are automatically inherited) and the `geom` to represent the model fit. Since the model fit lines don't have a point shape, we can also map Group to the linetype aesthetic so make the lines different between groups. We can get the fitted values for a `lmer` model object by using the `fitted` function.

```
> ggplot(amant.ex, aes(Week, DRS, shape=Group)) +
    stat_summary(fun.data=mean_se, geom="pointrange") +
    stat_summary(aes(y=fitted(m.1), linetype=Group),
                 fun.y=mean, geom="line")
```

The model fit in Figure 2.3 appears to be quite good and it specifically seems to capture the difference in rate of recovery between the placebo and amantadine groups. For both groups, the first and last data points are slightly above the line while the middle three points are slightly below it. This suggests that there might be a small amount of curvature in the recovery pattern – faster initial recovery that slows down over the course of the 4 weeks of treatment. The next chapter will discuss how to model more complex data curves.

2.4 Another example: Visual search response times

The primary focus of growth curve analysis – and longitudinal data analysis techniques in general – is to capture change over time, so most of the examples in this book will have some version of time as the main predictor. However, many of the same issues hold for other sorts of repeated measures scenarios with a continuous predictor. For example, response times in a conjunction visual search task tend to be a linear function of the number of objects in the display (called "set size"). In other words, when looking for a green letter "N" among green "X's" and brown "N's," response times are a linear function of the number of letters on the screen (e.g., Treisman & Gelade, 1980). The example data set `VisualSearchEx` contains results from a study that tested 15 participants with aphasia and 18 control participants in this task. We can treat set size the same way that we treat time and evaluate how response time changes as a function of set size for the two groups.

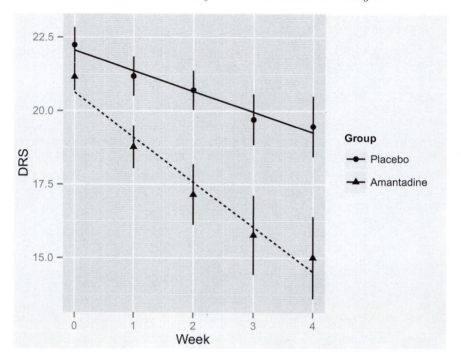

FIGURE 2.3
Observed data (symbols, vertical lines indicate ±SE) and linear model fits for recovery as measured by reduction in Disability Rating Scale (DRS) score over the four weeks of the study for the placebo and amantadine groups.

```
> summary(VisualSearchEx)
  Participant        Dx          Set.Size          RT
  0042   :  4   Aphasic:60   Min.   : 1.0   Min.    :   414
  0044   :  4   Control:72   1st Qu.: 4.0   1st Qu.:  1132
  0083   :  4                Median :10.0   Median :  1814
  0166   :  4                Mean   :12.8   Mean    :  2261
  0186   :  4                3rd Qu.:18.8   3rd Qu.:  2808
  0190   :  4                Max.   :30.0   Max.    : 12201
  (Other):108
```

The four variables in the data frame are

- **Participant**: a numeric identifier for each unique participant in the study (coded categorically).

- Dx: a categorical group factor ("Dx" is an abbreviation for "Diagnosis").

- **Set.Size**: the number of objects in the display (1, 5, 15, or 30).

- RT: average response time for each participant at each set size.

We build up the model the same way that we did for the amantadine example. Start with a base model of response times (RT) as a function of set size (Set.Size) and with individual participants varying in intercept and effect of set size:

```
> m.base <- lmer(RT ~ Set.Size + (Set.Size | Participant),
            data=VisualSearchEx, REML=FALSE)
```

Then we add the effects of diagnosis on the intercept and linear terms:

```
> m.0 <- lmer(RT ~ Set.Size + Dx + (Set.Size | Participant),
          data=VisualSearchEx, REML=FALSE)
> m <- lmer(RT ~ Set.Size * Dx + (Set.Size | Participant),
          data=VisualSearchEx, REML=FALSE)
```

Finally, we use the anova function to compare the models to evaluate the effect of diagnosis on response times:

```
> anova(m.base, m.0, m)
Data: VisualSearchEx
Models:
m.base: RT ~ Set.Size + (Set.Size | Participant)
m.0: RT ~ Set.Size + Dx + (Set.Size | Participant)
m: RT ~ Set.Size * Dx + (Set.Size | Participant)
        Df  AIC  BIC  logLik deviance Chisq Chi Df Pr(>Chisq)
m.base   6 2248 2265   -1118     2236
m.0      7 2241 2261   -1114     2227  8.58      1     0.0034 **
m        8 2241 2264   -1113     2225  2.01      1     0.1567
---
Signif. codes:  0 '***' 0.001 '**' 0.01 '*' 0.05 '.' 0.1 ' ' 1
```

and get the parameter estimates and their standard errors from the model summary:

```
> coef(summary(m))
                    Estimate Std. Error t value
(Intercept)         2078.749    264.361  7.8633
Set.Size              73.494     11.229  6.5449
DxControl          -1106.054    357.946 -3.0900
Set.Size:DxControl   -21.737     15.204 -1.4297
```

The analysis results indicate that there was a substantial group effect on the intercept and no effect on the slope. This can be seen in Figure 2.4: the aphasic group has much slower response times, but the two lines are essentially parallel (i.e., no difference in slope). Note that an effect on intercept and an effect on slope can have radically different theoretical implications: the intercept corresponds to a baseline or overall difference whereas the slope corresponds to a (linear) rate-of-change difference. One of the advantages of using this kind of regression approach is that it can go beyond just testing for differences – it can describe the *shape* of the differences.

Here is the code for generating Figure 2.4, which demonstrates the **error-bar** geom and some customization options.

```
> ggplot(VisualSearchEx, aes(Set.Size, RT,
                        shape=Dx, linetype=Dx)) +
    stat_summary(fun.y=mean, geom="point") +
    stat_summary(fun.data=mean_se, geom="errorbar",
              linetype="solid", width=0.6) +
    stat_summary(aes(y=fitted(m)), fun.y=mean, geom="line") +
    scale_shape_manual(values=c(1, 2)) +
    labs(x="Set Size", y="Response Time (ms)",
        linetype="Group", shape="Group") +
    theme_bw(base_size=10) +
    theme(legend.justification=c(0,1), legend.position=c(0,1),
        legend.background=
                element_rect(fill="white", color="black"))
```

The first line uses the **ggplot** command to set up the mappings of the data variables to different graph properties, including mapping **Dx** to both point shape (for the observed data) and linetype (for the model fits). There are three different data summaries: means of the observed data realized as points, means and standard errors of the observed data realized as errorbars and customized to have narrower crossbars (**width** argument) and to stop them from inheriting the linetype mapping (i.e., without this override, the Control group errorbars would be dashed), and mean of the model fit realized by lines. Specific point shapes are selected using **scale_shape_manual**, which overrides the default shape scale (the same approach works for defining other scales like color, **scale_color_manual**, and linetype, **scale_linetype_manual**). The axis labels are specified using **labs** and legend labels are treated the same way, in keeping with the grammar of graphics approach. Legend position and appearance is specified using **theme**.

2.5   Chapter recap

This chapter provided a basic overview of the math behind growth curve analysis and how it addresses the challenges described in Chapter 1. Using regression provides a way to treat time as a continuous variable instead of distinct time bins and provides a way to explicitly model the nested structure of the data. This chapter also introduced the distinction between fixed effects and random (residual error) effects. Fixed effects are those factors that the analyst

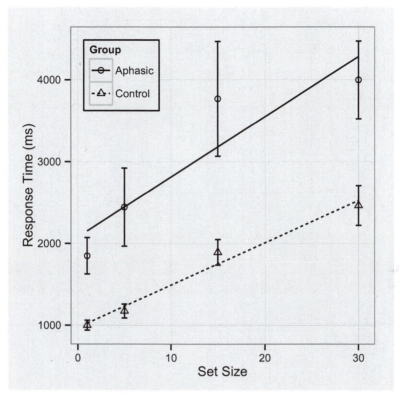

FIGURE 2.4
Conjunction visual search response times by participant group with linear
model fit lines.

believes to be reproducible, fixed properties of the world, and their parameters are estimated independently. Random effects correspond to observational
units that the analyst believes to be random samples from some population
to which (s)he wishes to generalize. Random effects are constrained to come
from a normal distribution with a mean of 0, which is meant to capture the
assumption that they reflect random samples and makes them interdependent.
Because they correspond to individual observational units, random effects also
capture the nested structure of the data and provide a way to quantify individual differences, which will be discussed in more detail in Chapter 7.

This chapter also provided the first two concrete examples of growth curve
analysis. We covered the core analysis syntax using the `lmer` function, how to
conduct model comparisons using the `anova` function, and how to plot model
fits using `ggplot`. The examples in this chapter constitute a starting point
– simple linear cases. The next chapters will build on this basic structure to
handle more complex situations.

2.6   Exercises

The `wisqars.suicide` data frame contains annual suicide rate data by state from 1999 to 2007 collected from the Web-based Injury Statistics Query and Reporting System (WISQARS) hosted by the Centers for Disease Control and Prevention. Use these data to analyze trends in suicide rates by geographic region (tip: adjust the `Year` variable to treat 1999 as time 0 so that it corresponds to the intercept).

1. Did the overall suicide rate in the US increase during this time period? If yes, what was the estimated rate of change?

2. Did the regions differ in their initial (1999) suicide rates?

3. Did the regions differ in their rate of change of suicide rate during this period?

4. Plot the observed and model-fit suicide rates by region for this time period. Include an indicator of variability (e.g., standard error) for the observed data. Make color and black-and-white versions.

3

When change over time is not linear

CONTENTS

3.1   Chapter overview

The previous chapter provided a conceptual overview of growth curve analysis and simple linear examples. Of course, time course data in the behavioral, cognitive, and neural sciences are rarely straight lines. Typically, the data have complex curved shapes, which means that the Level 1 model must also have a curved shape. The choice of the Level 1 model defines a *functional form* for the data; that is, the overall function or shape that will be used to describe the group and individual data. This choice is very important because it defines the framework for the whole analysis, so this chapter will describe some options and factors involved in choosing a functional form with a focus on one particularly good option: *higher-order polynomials*. This approach will be demonstrated with a step-by-step walk through a complete example, including how to estimate parameter-specific p-values and how to report growth curve analysis results.

3.2   Choosing a functional form

There are many different functional forms that one could potentially use to model data. This section will discuss the three major considerations involved in choosing a functional form for the kinds of data we typically encounter in the behavioral sciences.

3.2.1   Function must be adequate for the shape of the data

The first and perhaps most obvious consideration is that the functional form must be adequate for the shape of the data. For a functional form to be "adequate," it needs to be able to produce the shape of the observed data. Consider the curvilinear data in the left panel of Figure 3.1. A straight line (dotted line) clearly misses important aspects of the data, so it would not be adequate. The dashed curve does much better: it captures the overall approximately U-shaped pattern, though it seems to have some small — but systematic — deviations from the data. The solid curve fits the data even better. Keep in mind that it is neither possible nor advisable to try to capture every little blip in the data; doing so would *overfit* the data by forcing the model to describe the noise in the data as well as the effects of interest.

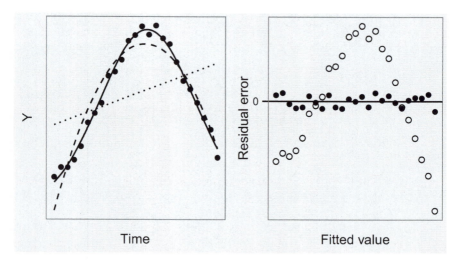

FIGURE 3.1

Left: Some hypothetical curvilinear data (black symbols) and three possible model fits. Right: Residual errors plotted against fitted values to examine systematic model deviations.

Distinguishing noise from potentially interesting effects is not always easy. A key difference is that noise is, by definition, not systematic. So if there are systematic deviations of the observed data from the model fits, that suggests that the functional form may not be adequate. It is sometimes easier to see such systematic deviations by plotting the residual error (i.e., deviation) against the model-predicted (fitted) values. This kind of plot shows what was missed by the model. For example, the right panel of Figure 3.1 shows this kind of plot for the worst and best models from the left panel. The residual errors from the linear model (open circles) start out negative, progressively become positive, then drop into the negatives again, which reflects the fact that the linear model missed the U-shape in the data. The residual errors from the non-linear model (filled circles) are not just closer to 0, they are also unsystematic – sometimes a little above 0, sometimes a little below, without any clear pattern. Systematic deviations can be more formally described in terms of *autocorrelated residuals*, which can be used to quantitatively assess whether a model is inadequate.

3.2.2   Dynamic consistency

When dealing with multilevel or nested data, we want the model to describe the overall average pattern as well as the individual participants' deviation from that average. Since the overall average pattern is the average of the individual participants, we want it to be true that if you fit the model to each participant's data individually and average together those parameter estimates, you would get the same values as if you had averaged together the data from all of the participants and fit the model to that average. This property is called *dynamic consistency*: the model of the average data is equal to the average of the models of individual participants' data.

Higher-order polynomials comprise one family of functions with this property. The term *higher-order* polynomials refers to polynomial functions of order greater than 1; for example, $Time^2$, $Time^3$, $Time^4$, etc. (the *order* or *degree* of a polynomial function is the value of the largest power or exponent in the function). These shapes are shown in Figure 3.2, where each panel shows the shape of a single polynomial time term. A growth curve model using higher-order polynomials would include multiple polynomial terms. For example, a second-order polynomial Level 1 model would have three β parameters:

$$Y_{ij} = \beta_{0i} + \beta_{1i} \cdot Time_j + \beta_{2i} \cdot Time_j^2 + \varepsilon_{ij} \qquad (3.1)$$

which correspond to the three time terms: the intercept (β_{0i}), the linear slope (β_{1i}), and the steepness of the quadratic curvature (β_{2i}). The Level 2 models of those parameters could then include condition (or other) effects on any or all of those Level 1 parameters, just as described in the previous chapter.

As shown in Figure 3.2, polynomial functions capture non-linear change over time, but if we consider $Time^2$, $Time^3$, etc., to be different predictors, then the model is just like a standard multiple linear regression model with different parameters for different predictors. In other words, polynomial functions are *non-linear in their variables* (i.e., Time), but *linear in their param-*

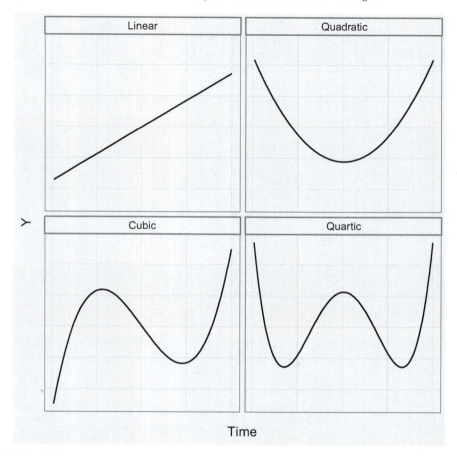

FIGURE 3.2
Schematic diagram of polynomial curve shapes. Each panel shows the shape
of just one polynomial term; a polynomial model fit can include a combination
of multiple terms.

eters. We can understand this distinction by considering functions that are
non-linear in both their variables and their parameters. One example is the
logistic power peak (LPP) function defined by Equation 3.2, which describes
the outcome variable Y at time j, in terms of curve amplitude α, curve width
β, peak location δ, and curve symmetry γ (for convenience the function is
written in two parts, with Equation 3.2a defining the overall LPP function
and Equation 3.2b defining a chunk that occurs twice in the overall function):

$$Y_j = \alpha \, (1 + \tau)^{\frac{-\gamma - 1}{\gamma}} \, \tau(\gamma + 1)^{\frac{\gamma + 1}{\gamma}} \tag{3.2a}$$

$$\tau = \exp\left(\frac{Time_j + \beta \ln(\gamma) - \delta}{\beta}\right) \tag{3.2b}$$

Unlike a polynomial function (e.g., Equation 3.1), the LPP function does not look like a multiple linear regression equation. Its non-linearity is not just a matter of having non-linear $Time$ predictor(s), the variables also have non-linear properties. Let's take a closer look at what happens if we try to use this function to model multilevel data.

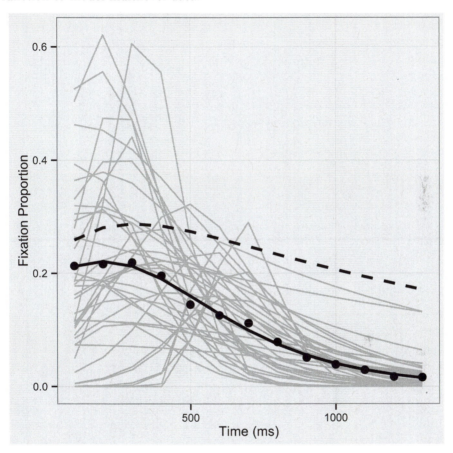

FIGURE 3.3

Consequences of lack of dynamic consistency in the logistic power peak function. The circles are the average of the observed data. The solid black line is the model of those average data. The grey lines are the models of the individual participants. The dashed line is the curve produced by averaging the parameters from the models of the individual participants. The dashed line does not match the solid line, nor does it reflect the central tendency of the models of the individual participants.

The data for this example come from an eye-tracking experiment (Mirman & Magnuson, 2009) in which the outcome measure was the probability of looks ("fixation proportion") to a particular object at each point in time. The black circles in Figure 3.3 show the overall average of the observed data from one of the conditions in the experiment and the solid black line is the LPP model fit to those average data. The LPP function has been used to analyze this sort of fixation time course data (e.g., Scheepers, Keller, & Lapata, 2008) and clearly fits the data very well. The grey lines in Figure 3.3 are the models of data from individual participants (there were 38 participants) and these also fit the individual data very well. However, when the parameters (α, β, δ, and γ) from those individual participant models were averaged together, the resulting model (dashed line) was substantially different from both the model of the average data (solid black line) and the observed data (symbols). It also did not reflect the central tendency of the individual participant models (grey lines), which is most obvious in the later half of the time course, where the dashed line is above all of the grey lines; that is, the average of the individual models is higher than any of the individual models.

This mismatch between the central tendency of the individual models and the average of the individual models poses quite serious problems. The biggest one is that it makes standard inferential statistical methods useless. For example, one might want to use a paired-samples t-test to evaluate whether the α (amplitude) parameter was different between two conditions in the experiment. The problem is that this test would be using the individual participants' α parameters to evaluate whether the *means* for the two conditions were different, but those means would no longer reflect a model of the actual fixation data (i.e., those means would represent the dashed line in Figure 3.3, which reflects neither the average data nor the individual models). Even if one used other, more sophisticated model or parameter evaluation techniques, they would necessarily be based on estimating means and variances from individual observations, but those means and variances are linear properties and the dynamically inconsistent LPP functional form has only non-linear properties. (For another discussion of the problems that arise when the model of the average does not reflect the individual data see Brown & Heathcote, 2003.)

A corollary of this problem is that lack of dynamic consistency makes any kind of individual differences analysis uninterpretable. For example, one might want to test a correlation between individual participants' α parameters and scores on some other cognitive test. But, again, there is the problem that the average of the individual parameters does not reflect the central tendency of the individual models. In other words, it would mean running a linear correlation using a non-linear variable. Even worse, the particular non-linearity of the variable is very opaque, so it is hard to know what kind of artifacts it will introduce into the correlation analysis.

Finally, recall from Chapter 2 that in a multilevel regression framework, the random effect parameters are constrained to be drawn from a normal distribution with a mean of 0. That is, the average of the individual deviations (random effects) from the overall mean (fixed effects) is 0. Which is to say that the average of the individual models is equal to the model of the average. This means that dynamic consistency is not just an important statistical property, it is *required* for multilevel regression.

Logistic regression is one important class of regression models that might, at first glance, appear not to be dynamically consistent. In fact, they are dynamically consistent and the easiest way to understand why is to think about what parameters will be estimated by the model. In a polynomial logistic regression, the estimated parameters will be polynomial coefficients, which are dynamically consistent. In a non-linear regression model with a logistic functional form, the estimated parameters are the parameters of the logistic function, which are generally not dynamically consistent.[1] Chapter 6 will describe why logistic regression is important and how to use logistic GCA.

3.2.3 Making predictions: Fits and forecasts

Models can make two fundamentally different kinds of "predictions": model fits within the boundaries of the observed data and model forecasts for what would happen outside of those boundaries. For forecast-type predictions it is critical that the model have a functional form that matches the form of the system that generated the observed data. If it does not, then its forecasts are likely to be wrong. This is precisely the logic of cognitive or computational modeling (as opposed to statistical modeling): build a model that has the hypothesized properties of the system under investigation and conduct simulations of that model to see if it fits the observed data. Critically, making novel predictions — that is, forecasts for what should happen outside the boundaries of the data that were used to develop the model — is considered the best test of computational models (for a good discussion of computational modeling in cognitive science see McClelland, 2009). Computational models are meant to implement theories and theories need to be falsifiable, so making and testing falsifiable predictions is a critical aspect of computational modeling. When a computational model's prediction is found to be wrong, this provides critical evidence that the model was wrong, either in theory or in implementation, and is grounds for rejecting that model or at least preferring an alternative (for more discussion of computational model evaluation and rejection see Magnuson, Mirman, & Harris, 2012).

Fit-type predictions are more accurately called *quantitative descriptions* of the observed data because their goal is to summarize the data patterns rather than to make novel (falsifiable) predictions. It is useful to have a description of the data that is not tied to a particular theory because the data are always "true," but a theory can be wrong. For example, imagine that we made some very careful neuroanatomical measurements but then described them entirely within the framework of phrenology (e.g., sizes of the organ of friendship, the organ of courage, etc.). The measurements could have remained informative

---

[1]It may be possible to constrain the logistic function to be dynamically consistent, but doing so is quite difficult and requires fairly complex programming, so it will not be covered in this book.

if they had been in theory-neutral units (like millimeters), but using phreno-logical units meant that the measurements became useless once the theory of phrenology was discredited. Returning to statistical modeling, let's say we have some data on the relationship between physical stimulus magnitude and its perceived intensity. If we assume that Weber's Law holds in this domain, then stimulus magnitude and perceived intensity should have a power law relationship, so we can fit a power law function and compute the Weber constant for this domain. However, if it happens that Weber's Law does not hold in this domain, then the Weber constant is not a useful description of the data. Had we described the data in a theory-neutral way, we would be able to evaluate how well any theory accounts for the data, not just Weber's Law.

It is also useful to have a *quantitative* description of the observed data because it allows quantitative comparisons with predictions from different theories. A qualitative description such as "perceived intensity increases with increasing physical stimulus magnitude" is a reasonable starting point, but quickly becomes limited in terms of distinguishing between different accounts. When the full set of the observed data are available for analysis, one can fit multiple theoretical models directly to the observed data in order to obtain a quantitative evaluation (e.g., Oberauer & Kliegl, 2006; Wagenmakers & Farrell, 2004). However, that evaluation would be limited to the particular models (as discussed above), so it would be necessary to make all observed data publicly available so that any future proposed theory could be compared to every data set. This kind of data sharing is not the norm (though many argue that it should be), so there remains a need for researchers to describe their findings in a way that is formal enough to allow quantitative comparisons with different theoretical models and independent of specific theories so the data remain relevant as theories evolve and new theories emerge.

The bottom line is that computational and statistical models serve distinct and complementary roles. Statistical models provide descriptions of large data sets in terms of a small set of effects or patterns and quantify those effects or patterns in ways that can be compared against any theoretical account. Computational models instantiate a particular theory in a way that allows concrete testing and making predictions. Because they serve complementary roles, statistical and computational models can be combined to form a powerful two-pronged research strategy: using statistical models to describe the data and using computational models to evaluate theories against those descriptions (for examples of such a two-pronged strategy in the domain of spoken word recognition see Mirman, Dixon, & Magnuson, 2008; Mirman et al., 2011).

3.3   Using higher-order polynomials

3.3.1 Strengths and weaknesses

Higher-order polynomial functions satisfy the constraints described above. They are not the only family of functions that do so and other functional forms can be used for growth curve analysis (e.g., Cudeck & Harring, 2007; Grimm, Ram, & Hamagami, 2011; Oberauer & Kliegl, 2006; Pinheiro & Bates, 2000), but implementing these other models in `lme4` is rather challenging (Bolker, 2013; Kliegl, 2013). This book will focus on polynomial GCA because it is comparatively easy to implement and effective for most cases, but most of the content applies to GCA regardless of the specific functional form. Polynomial functions are dynamically consistent and a polynomial of sufficiently high order is guaranteed to provide an arbitrarily good fit to the observed data (this is known as *Taylor's Theorem*). That is, for any set of observed data, polynomials can describe those data in a dynamically consistent way and as accurately as desired, if you just add enough polynomial terms ($Time^2$, $Time^3$, $Time^4$, etc.).

This is not to say that polynomial functions are without weaknesses. Polynomial functions are not *asymptotic* – they don't have flat plateau-like sections (see Figure 3.2), so they can have difficulty fitting asymptotic data. For example, a weight loss study might show that weight initially decreases and then stabilizes at a final level. As shown by Taylor's Theorem, polynomials *can* capture this pattern, but it is not their intrinsic form. This difficulty may be (partly) resolved by restricting the length of the tail data that are included in the analysis: if by the 5th week participants are already very close to their final weight and there is no change over the course of the next 5 weeks, perhaps the analysis could just focus on the first 5 or 6 weeks. It is important not to introduce experimenter bias by doing this kind of truncation. Ideally, the researcher can use an unbiased method to define the time window for the analysis. If that is not available, an alternative strategy is to conduct parallel analyses using a few different time windows (e.g., 4 weeks, 5 weeks, 6 weeks, etc.) and verify that the pattern of results is generally the same, so the scientific claims will not depend too strongly on the specific analysis window.

It is also useful to consider *why* the data are asymptotic – asymptotes can arise because the underlying process actually plateaued or due to floor or ceiling effects. For example, proportions are bounded at 0 and 1, and tend to plateau as they approach those bounds. In such cases it is better to use logistic or quasi-logistic regression (described in Chapter 6) with a polynomial functional form rather than non-linear functional forms.

More generally, it is important to keep in mind that the cognitive, neural, developmental, etc., processes under investigation are probably not polynomial, so polynomials are a poor choice for forecast-type predictions. As discussed in section 3.2.3, this sort of prediction is generally the realm of computational or theoretical models, not of statistical models. The statistical model's goal is to provide a quantitative description of the data that can be used to evaluate any theoretical model. As such, the positive statistical and implementational properties of higher-order polynomials outweigh their negative properties, but it is important to remember that polynomial functions may be good at describing the patterns within the boundaries of the observed data and should not be used to make predictions for what will happen outside those boundaries.

3.3.2   Choosing polynomial order

Using higher-order polynomials requires choosing the specific order of the polynomial to be used for each analysis. Is a quadratic function sufficient or does it need to be cubic, or quartic, etc.? As a general guideline, it can be helpful to think about the data in terms of how many times the curve changes direction (more formally, this corresponds to the number of *inflection points*). A flat line has zero changes of direction, so that could be modeled with a zero-order polynomial (just the intercept: β_0). A straight, non-flat line, has an initial change of direction from flatness, so that could be modeled with a first-order polynomial ($\beta_0 + \beta_1 Time$). A U-shaped curve has the initial change from flatness and the reversal at the bottom of the U, so that could be modeled by a second-order polynomial ($\beta_0 + \beta_1 Time + \beta_2 Time^2$), and so on. This is just a general starting point for generating intuitions about the appropriate polynomial order for a particular data shape. There are a few different, more formal approaches and it is probably best to consider all of them without being too doctrinaire about any of them.

A *statistical* approach is to include only and all of the polynomial orders that improve model fit. Following this approach, you could start with a simple linear Level 1 model, then add a quadratic term and test whether it improves model fit, then a cubic term, and so on, until the additions no longer improve model fit. This approach has the advantage of statistical rigor, but it can lead to the inclusion of uninterpretable effects.

A *theoretical* approach takes the complementary view: the model should include only those terms for which the experimenter predicted an effect. This approach is guaranteed to produce only interpretable results because only predicted effects would be tested, but it can miss important unpredicted effects if the model does not adequately capture the data.

One way to combine these two approaches is to use the statistical approach to define the Level 1 model and the theoretical approach to constrain which of the Level 2 models will have effects of manipulations. For example, if an experimenter predicted that some manipulation would produce a simple baseline shift in some complex behavior, then the overall (Level 1) data shape might require a high-order polynomial, but the condition (i.e., experimental manipulation) effect could be included only in the Level 2 model of the intercept term.

It is also useful to keep practical considerations in mind. Most common data shapes can be captured with 4th-order or lower polynomials and terms beyond the quadratic (2nd-order) become progressively more difficult to interpret. With that in mind, it is rarely advisable to go beyond 4th-order polynomials.

3.3.3 Orthogonal polynomials

For higher-order polynomials, the individual time terms tend to be correlated. For example, as the linear time term increases, so does the quadratic time term. This kind of colinearity among predictors means that their parameter estimates cannot be evaluated independently because the predictors are trying to capture some of the same variance in the data. Colinearity can also undermine the stability of parameter estimates: small changes in the data, such as the addition or deletion of a single data point or even just inherent measurement noise, can lead to very large changes in the parameter estimates. A relatively standard solution to the colinearity problem is to orthogonalize the predictors, which removes their correlation. In the case of polynomials, this means creating *orthogonal polynomials*.

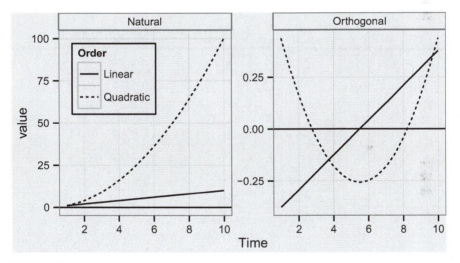

FIGURE 3.4
Examples of natural (left) and orthogonal (right) linear and quadratic polynomials.

Figure 3.4 shows first-order (linear) and second-order (quadratic) polynomial time functions with the natural polynomial version on the left and the orthogonal polynomial version on the right. In both versions, the linear term captures linear change and the quadratic captures U-shaped parabolic change, but in the orthogonal version the two time terms are centered and scaled to be in the same range of outcome values. The scaling is convenient because it means that their parameter estimates will be on the same scale. More importantly, the centering makes the two time terms uncorrelated over the range of time values (that is, *orthogonal*), which means that their parameter estimates will be independent.

Orthogonal polynomials are just a useful transformation of natural polynomials. This transformation requires a specified range (time window) and order. The range needs to be specified because the centering is specific to the time window – the particular orthogonal polynomials in Figure 3.4 would no longer be uncorrelated if they were extended to time values of 20 or -10. The order needs to be specified because all of the terms need to be orthogonal to all of the other terms. Conveniently, the R function `poly` will create an orthogonal polynomial for a specified range and order. Practically, the main implication of this is that one needs to choose the range and order before running the analysis and changing the range or order (for example, analyzing a smaller time window) requires re-starting from the initial step of creating the orthogonal polynomial time terms.

Natural and orthogonal polynomial terms have the same shapes, but the centering of orthogonal polynomials gives them slightly different interpretations compared to natural polynomials. A particularly important difference concerns the intercept term. For natural polynomials, the intercept term corresponds to the *y*-intercept; that is, the outcome value when the predictor value is 0. In many cases, this is a baseline value, which may be of special theoretical importance. For example, in the the previous chapter's brain injury recovery study example, the fact that there was no significant group effect on the intercept indicated that the two groups were (approximately) equally severely impaired before administration of the drug. This is important because it means that the differences in recovery rate cannot be attributed to initial severity. Conversely, studies that examine whether listeners use context information to predict or anticipate words during language comprehension may specifically predict baseline differences (e.g., Barr, Gann, & Pierce, 2011).

For orthogonal polynomials, the intercept term corresponds to the overall average, which can also be a useful measure. To concretely demonstrate the difference, Figure 3.5 shows the brain injury recovery example from the previous chapter, with arrows indicating the natural and orthogonal intercepts. The model predictions for the overall pattern are exactly the same for natural and orthogonal versions, but the intercept terms correspond to different aspects of the data. In fact, the estimated group effect on the intercept is more than twice as large for the orthogonal version than the natural version (3.09 vs. 1.43) because it reflects the average difference in severity between the two groups across the whole duration of treatment, not just at the starting point.

In the brain injury recovery example it was important to know the *y*-intercept, which made natural polynomials a more useful approach. In other cases knowing the overall average (or "area under the curve") is more useful. For example, in the visual search example from the previous chapter, the orthogonal intercept would estimate the overall mean reaction time difference between the control and aphasic groups, which may be more informative than

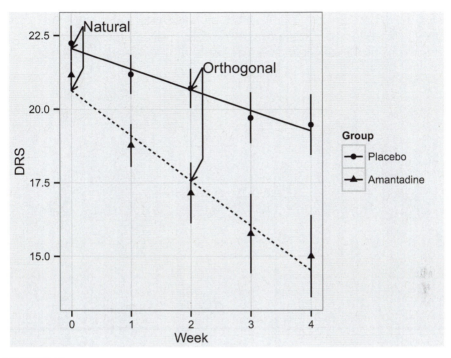

FIGURE 3.5
Natural and orthogonal intercepts in data from a study of the effect of amantadine on recovery from brain injury.

the expected reaction time difference at set size 0 (i.e., when there are 0 objects
on the screen).

The bottom line is that natural polynomials allow testing for differences
at "Time 0," so the natural polynomial approach is more useful when such
differences need to be tested. Otherwise, orthogonal polynomials are generally
better because the time terms are independent. The next section will discuss
interpreting higher-order polynomial effects in more detail.

3.3.4   Interpreting higher-order polynomial effects

To interpret polynomial effects in the context of complex data shapes it can
be useful to think of each term as a separate component for the observed data
curve. This is particularly useful for orthogonal polynomials because those
components are necessarily independent. Figure 3.6 shows how each (orthogonal) polynomial shape changes with changes of its coefficient. As coefficients
move toward 0, the function approaches a flat line; as the coefficients become larger, the function becomes steeper. For negative coefficients, the shapes
would just be inverted (upside-down). "Steeper" means different things for the
different components: for the linear component, this is the familiar steeper

ramp; for the quadratic component it is the sharpness of the (centered) peak; similarly, for the cubic and quartic it is the sharpness of the peaks (two peaks for the cubic, three peaks for the quartic).

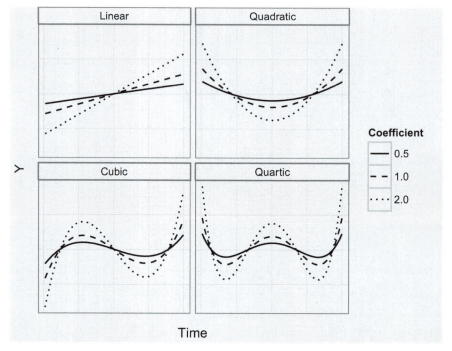

FIGURE 3.6
Schematic examples of orthogonal polynomial terms with different coefficients.

Understanding the polynomial components is a critical first step toward understanding polynomial effects in a growth curve analysis – these are the pieces that are being combined and manipulated when one fits a growth curve model to real data. Of course, those analyses will involve multiple components combined in complex ways, so understanding the components is only the first step. A useful second step is visually comparing fits from models with and without particular components to see how they differ. The next section will walk through a step-by-step example of growth curve analysis using orthogonal polynomials; the next chapter will include an example of plotting different model fits to interpret effects of high-order polynomial terms.

3.4 Example: Word learning

In Chapter 1, we saw that traditional *t*-test and ANOVA approaches were not effective at capturing the effect of transitional probability (TP) on the rate of novel word learning. These example data are taken from a real experiment (Mirman, Magnuson, Graf Estes, & Dixon, 2008) and reproduced in Figure 3.7. Let's analyze them using GCA. The first step should always be to look at the data, both in text form and graphically.

```
> summary(WordLearnEx)
    Subject         TP            Block            Accuracy
 244    : 10   Low :280    Min.   : 1.0    Min.   :0.000
 253    : 10   High:280    1st Qu.: 3.0    1st Qu.:0.667
 302    : 10               Median : 5.5    Median :0.833
 303    : 10               Mean   : 5.5    Mean   :0.805
 305    : 10               3rd Qu.: 8.0    3rd Qu.:1.000
 306    : 10               Max.   :10.0    Max.   :1.000
 (Other):500
```

The data frame contains 4 variables:

- `Subject`: A unique identifier for each participant. The identifier is numeric, but treated as a categorical factor. The summary tells us that there are 10 observations per participant.

- `TP`: A categorical between-participants factor with two levels, low and high (within-participants manipulations will be covered in Chapter 4). There are 280 observations in each condition, 10 for each of 28 participants.

- `Block`: A numeric variable indicating training block, ranging from 1 to 10.

- `Accuracy`: Proportion correct for a given participant in a given training block, ranging from 0 to 1.

Here is the code for generating Figure 3.7:

```
> ggplot(WordLearnEx, aes(Block, Accuracy, shape=TP)) +
    stat_summary(fun.y=mean, geom="line", size=1) +
    stat_summary(fun.data=mean_se, geom="pointrange", size=1) +
    theme_bw(base_size=10) +
    coord_cartesian(ylim=c(0.5, 1.0)) +
    scale_x_continuous(breaks=1:10)
```

For data like these, a second-order polynomial should suffice. We'll use orthogonal polynomials for a few reasons. First, in the experiment, participants

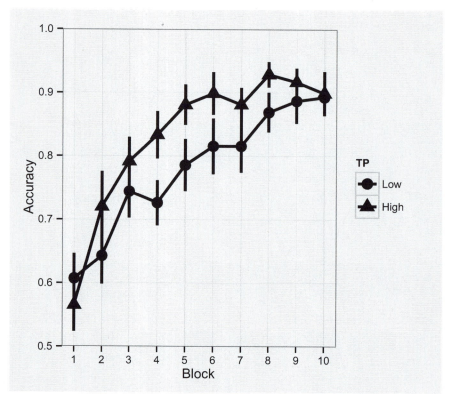

FIGURE 3.7
Effect of transitional probability (TP) on novel word learning.

learned to match a made-up spoken "word" like *pibu* with a novel geometric shape. All of these "words" were completely novel and arbitrarily paired with shapes and counterbalanced across participants. There were two shape choices on each trial, so it is not very interesting that accuracy would start around 50%, making the y-intercept not very informative. On the other hand, the overall mean accuracy does (partially) reflect faster learning, so the orthogonal intercept will be more informative. Second, orthogonal polynomials will make the linear and quadratic terms uncorrelated, so we will be able to independently evaluate the linear slope and the steepness of the curvature. We can use the `poly` function to create a second-order orthogonal polynomial in the range of `Block`:

```
> t <- poly(unique(WordLearnEx$Block), 2)
```

Now we need to add those orthogonal polynomial values into the original data frame aligned by `Block`. The following command will do that by creating two new variables, `ot1` and `ot2` (for <u>o</u>rthogonal <u>t</u>ime order <u>1</u> and <u>o</u>rthogonal <u>t</u>ime order **2**), in the `WordLearnEx` data frame and using `Block` as an index in the orthogonal polynomial variable `t`:

```
> WordLearnEx[,paste("ot", 1:2, sep="")] <-
                         t[WordLearnEx$Block, 1:2]
```

We can re-check the data frame and see that the summary now shows the two new variables:

```
> summary(WordLearnEx)
    Subject        TP            Block          Accuracy
 244    : 10   Low :280   Min.    : 1.0   Min.    :0.000
 253    : 10   High:280   1st Qu.: 3.0    1st Qu.:0.667
 302    : 10              Median : 5.5    Median :0.833
 303    : 10              Mean    : 5.5   Mean    :0.805
 305    : 10              3rd Qu.: 8.0    3rd Qu.:1.000
 306    : 10              Max.    :10.0   Max.    :1.000
 (Other):500
       ot1                 ot2
 Min.    :-0.495   Min.    :-0.348
 1st Qu.:-0.275    1st Qu.:-0.261
 Median : 0.000    Median :-0.087
 Mean    : 0.000   Mean    : 0.000
 3rd Qu.: 0.275    3rd Qu.: 0.174
 Max.    : 0.495   Max.    : 0.522
```

Now we begin the analysis with a base model that just has the Level 1 structure and the random effects, without any effects of TP:

```
> m.base <- lmer(Accuracy ~ (ot1+ot2) + (ot1+ot2 | Subject),
              data=WordLearnEx, REML=FALSE)
```

This base model just captures the overall time course with a second-order orthogonal polynomial and allows individual participants to vary randomly (i.e., following a normal distribution with a mean of 0) on any of the three components of the overall time course (intercept, linear, and quadratic). Now we can add the experimental effects, starting with the fixed effect of TP on the intercept:

```
> m.0 <- lmer(Accuracy ~ (ot1+ot2) + TP + (ot1+ot2 | Subject),
            data=WordLearnEx, REML=FALSE)
```

then also on the linear term:

```
> m.1 <- lmer(Accuracy ~ (ot1+ot2) + TP + ot1:TP +
                   (ot1+ot2 | Subject),
              data=WordLearnEx, REML=FALSE)
```

and finally the full model with effects of TP on all time terms, written in a more compact form:

```
> m.2 <- lmer(Accuracy ~ (ot1+ot2)*TP + (ot1+ot2 | Subject),
              data=WordLearnEx, REML=FALSE)
```

Now we can evaluate the effect of adding each term by using model comparisons:

```
> anova(m.base, m.0, m.1, m.2)
Data: WordLearnEx
Models:
m.base: Accuracy ~ (ot1 + ot2) + (ot1 + ot2 | Subject)
m.0: Accuracy ~ (ot1 + ot2) + TP + (ot1 + ot2 | Subject)
m.1: Accuracy ~ (ot1 + ot2) + TP + ot1:TP +
              (ot1 + ot2 | Subject)
m.2: Accuracy ~ (ot1 + ot2) * TP + (ot1 + ot2 | Subject)
       Df  AIC  BIC logLik deviance Chisq Chi Df Pr(>Chisq)
m.base 10 -331 -288    175     -351
m.0    11 -330 -283    176     -352  1.55      1      0.213
m.1    12 -329 -277    176     -353  0.36      1      0.550
m.2    13 -333 -276    179     -359  5.95      1      0.015 *
---
Signif. codes:  0 '***' 0.001 '**' 0.01 '*' 0.05 '.' 0.1 ' ' 1
```

The model comparison results show that the only effect of TP that significantly improved model fit was on the quadratic term. Figure 3.8 shows the behavioral data with the model fit, confirming that the model captured the faster novel word learning in the high TP condition.

```
> ggplot(WordLearnEx, aes(Block, Accuracy, shape=TP)) +
    stat_summary(aes(y=fitted(m.2), linetype=TP), fun.y=mean,
                 geom="line", size=1) +
    stat_summary(fun.data=mean_se,geom="pointrange",size=1)+
    theme_bw(base_size=10) +
    coord_cartesian(ylim=c(0.5, 1.0)) +
    scale_x_continuous(breaks=1:10)
```

3.5   Parameter-specific *p*-values

Model comparisons provide the best test of whether a particular effect made a statistically significant contribution to model fit and this approach should be used whenever possible. However, in some cases it may be valuable to evaluate

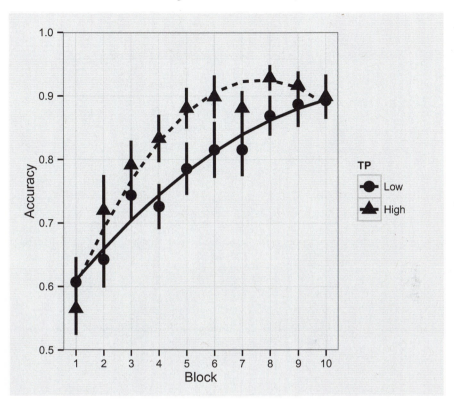

FIGURE 3.8
Observed data and growth curve model fits for effect of transitional probability
(TP) on novel word learning.

individual parameter estimates independent of overall model comparisons. The
parameter estimates constitute a measure of effect size, so they can be useful
in discussions of the clinical or practical significance (as opposed to statistical
significance) of an effect. Also, when a predictor has more than two discrete
conditions, the default behavior for `lmer` is to treat one condition as the
baseline and estimate parameters for each of the other conditions (Chapter 5
will discuss how to handle this sort of situation in more detail). The model
comparison will then include the effect of adding all of those parameters ($N-1$,
for N conditions) and it will be impossible to tell which of the conditions is
different from the baseline. In this case, a parameter-specific evaluation of
statistical significance is needed.

In principle, the parameter-specific evaluation is a one-sample t-test evaluating whether the estimated parameter is different from 0. The t-value corresponds to the parameter estimate divided by its standard error, but the
problem is that, for multilevel regression parameters, the degrees of freedom
for that t-test are not well-defined, so a standard t-test can't be done. One

option is to use Markov Chain Monte Carlo (MCMC) simulation to generate a 95% confidence interval for each parameter estimate (e.g., Baayen et al., 2008). However, this approach is limited in terms of the kinds of random effect structures that it can handle and the resulting p-values can be severely anticonservative. That is, $p < 0.05$ is meant to indicate a less than 5% probability of falsely detecting a non-existent effect, but the actual false alarm rate may be substantially higher (as high as 20% in some cases; for a detailed analysis see Barr et al., 2013).

A simple alternative is to use the normal distribution as an approximation. As the number of degrees of freedom increases, the t distribution converges to the normal distribution. This means that when the degrees of freedom are relatively large, the normal distribution can be used as an approximation. If the degrees of freedom are sufficiently large, then the specific number of degrees of freedom will have little effect. Concretely, the t distribution for 10 degrees of freedom differs substantially from the t distribution for 15 degrees of freedom but the t distributions for 510 and 515 degrees of freedom are essentially identical. Time course data typically involve a large number of observations relative to the number of fixed effect parameters in the model. For example, in the word learning example there were 560 observations (10 training blocks for each of 56 participants) and only 6 fixed effect parameters. Because the t distribution converges to the normal distribution when there are many degrees of freedom and because time course data typically have many observations, we can use the normal distribution to calculate approximate p-values. These approximations will be somewhat anticonservative, but some analyses suggest that the anticonservativity will not be too severe (e.g., a false alarm under 10% for $p < 0.05$; see Barr et al., 2013).

To use the normal approximation, we first turn the model's parameter estimates into a new data frame. The fixed effect parameter estimates can be extracted from the model summary using the `coefs` function:

```
> coefs <- data.frame(coef(summary(m.2)))
```

We then look up the p-value that corresponds to the absolute value of the t-value in the normal distribution (using the `pnorm` function), subtract it from 1 to get the probability of a t-value exceeding the observed value, multiply it by 2 to get a two-tailed p-value, and assign it to a new variable (called `p`) in the `coefs` data frame

```
> coefs$p <- 2 * (1 - pnorm(abs(coefs$t.value)))
```

We can now see the full set of fixed effect parameter estimates and their p-values:

```
> coefs
              Estimate  Std..Error   t.value           p
(Intercept)  0.7785250   0.021728  35.830648  0.0000e+00
ot1          0.2863155   0.037789   7.576772  3.5527e-14
```

```
ot2          -0.0508493   0.033188 -1.532182 1.2548e-01
TPHigh        0.0529607   0.030728  1.723538 8.4791e-02
ot1:TPHigh    0.0010754   0.053441  0.020123 9.8395e-01
ot2:TPHigh   -0.1164548   0.046934 -2.481234 1.3093e-02
```

The last three parameter estimates correspond to the effects of TP and these results are quite similar to the model comparison results: only the effect of TP on the quadratic term is statistically significant.

The `lmerTest` package offers a set of somewhat more sophisticated alternative approximations. The simplest approach is just to let the `lmerTest` package add parameter-specific p-values when fitting a model using `lmer`. These p-values are calculated using Satterthwaite's approximation for degrees of freedom. Here is an example using the full model of the word learning data:

Start by loading the package:

```
> library(lmerTest)
```

then refit the model:

```
> m.2t <- lmer(Accuracy ~ (ot1+ot2)*TP + (ot1+ot2 | Subject),
               data=WordLearnEx, REML=F)
```

The syntax is identical because it is the same `lmer` function, but now `lmerTest` has calculated parameter-specific p-values, which can be seen in the summary:

```
> coef(summary(m.2t))
              Estimate Std. Error     df   t value    Pr(>|t|)
(Intercept)  0.7785250   0.021728 56.008 35.830648 0.0000e+00
ot1          0.2863155   0.037789 62.507  7.576772 2.0520e-10
ot2         -0.0508493   0.033188 93.236 -1.532182 1.2886e-01
TPHigh       0.0529607   0.030728 56.008  1.723538 9.0308e-02
ot1:TPHigh   0.0010754   0.053441 62.507  0.020123 9.8401e-01
ot2:TPHigh  -0.1164548   0.046934 93.236 -2.481234 1.4885e-02
```

Notice that the parameter estimates, standard errors, and t-values are unchanged, because the model-fitting was identical, but the p-values are slightly different from the normal approximation because the estimated degrees of freedom are different.

3.6 Reporting growth curve analysis results

When describing study methods, the key principle is to provide enough information for another researcher to be able to replicate the study. This general principle extends to statistical analysis methods as well: provide enough information that another researcher would be able to replicate your analysis. Common analysis methods like *t*-tests and ANOVAs are so standardized that it can be enough to simply identify the method. More sophisticated methods like growth curve analysis have many possible variations, so it is important to provide all of the key details of the analysis method and results. There are three general categories of information that need to be included:

1. **The model structure.** The same general analysis framework can be applied in many different ways, so it is not enough to say that you used "multilevel modeling" or "growth curve analysis." It is important to clearly describe the functional form, all of the fixed effects, and the random effects structure.

2. **The basis for the inferential statistics.** Readers need to know how you made inferences about your model. That is, what method you used for significance testing. For model comparisons, make sure to clearly describe the models that were compared. For parameter-specific *p*-values report that the normal approximation was used.

3. **Complete model results, not just *p*-values.** For model comparisons, report the change in log-likelihood and the degrees of freedom (i.e., the χ^2 test). For parameter estimates, report the estimates and their standard errors (the *t*-values are optional because they are just the estimates divided by standard errors).

It may also be a good idea to include information about the software used to do the analysis. Here is an example of how these pieces might be combined for the word learning example from section 3.4:

> Growth curve analysis (Mirman, 2014) was used to analyze the learning of the novel words over the course of 10 training blocks. The overall learning curves were modeled with second-order orthogonal polynomials and fixed effects of TP on all time terms. The low TP condition was treated as the baseline and parameters were estimated for the high TP condition. The model also included random effects of participants on all time terms. The fixed effects of TP were added individually and their effects on model fit were evaluated using model comparisons. Improvements in model fit were evaluated using -2 times the change in log-likelihood, which is distributed as χ^2 with degrees of freedom equal to the number of parameters added. All analyses were carried out in R version 3.0.2 using the `lme4` package (version 1.0-5).
>
> The effect of TP on the intercept did not improve model fit ($\chi^2(1) = 1.55, p = 0.213$), nor did the effect of TP on the linear term ($\chi^2(1) = 0.358, p = 0.55$). The effect of TP on the quadratic term, however, did improve model fit ($\chi^2(1) = 5.95, p = 0.0147$),

indicating that the low and high TP conditions differed in the rate of word learning. Table 3.1 shows the fixed effect parameter estimates and their standard errors along with p-values estimated using the normal approximation for the t-values.

TABLE 3.1
Parameter Estimates for Analysis of Effect of TP on Novel Word Learning

	Estimate	Std. Error	t	p
Intercept	0.779	0.022	35.831	0.000
Linear	0.286	0.038	7.577	0.000
Quadratic	-0.051	0.033	-1.532	0.125
High TP: Intercept	0.053	0.031	1.724	0.085
High TP: Linear	0.001	0.053	0.020	0.984
High TP: Quadratic	-0.116	0.047	-2.481	0.013

3.7 Chapter recap

This chapter focused on extending the basic version of growth curve analysis beyond just straight lines. Modeling non-linear effects of time begins with choosing an overall function or shape to describe the data, which is called the *functional form*. Three principles for selecting a functional form were discussed: (1) it must be adequate to the data, (2) it must be dynamically consistent, and (3) it must be able to make the kinds of predictions that the analyst wants to make. Higher-order polynomials are one set of functional forms that satisfy these constraints, with some important caveats regarding asymptotic data and making forecast predictions. With this in mind, the chapter covered how to choose polynomial order and a useful transformation — orthogonal polynomials — that makes polynomial time terms independent.

Growth curve analysis using orthogonal polynomials was then demonstrated with an analysis of word learning data. These data had been discussed in Chapter 1 as an example of time course effects not captured by t-tests and ANOVAs. The example showed how to generate the needed orthogonal polynomials, analyze the data, interpret the results and plot the model fits. This chapter also discussed two methods for evaluating statistical significance of effects of interest. The better method is using model comparisons to evaluate improvement in model fit due to particular effects. When parameter-specific p-values are desired, the normal approximation provides a viable alternative that is only somewhat anticonservative. The last section discussed three principles for reporting growth curve analysis results: (1) report the full model structure, (2) report the basis for the inferential statistics, and (3) report complete model results.

3.8   Exercises

The CP data frame contains auditory discrimination data (d', called "d prime") for two continua of eight stimuli. The continua were created by morphing between two sounds from different categories, either along a temporal acoustic dimension or along a spectral acoustic dimension. The hypothesis was that there would be "categorical perception" — better discrimination near the category boundary than near the endpoints — for the temporal dimension but not for the spectral dimension (Mirman, Holt, & McClelland, 2004).

1. Analyze these data using growth curve analysis with second-order orthogonal polynomials. Which polynomial terms show statistically significant effects of continuum type?

2. Estimate parameter-specific p-values using the normal distribution. How does this evaluation of the effects of continuum type compare with the model comparisons approach? Repeat using `lmerTest`.

3. Plot the observed and model-fit d' by continuum. Include an indicator of variability (e.g., standard error) for the observed data. Make color and black-and-white versions.

4

Structuring random effects

CONTENTS

4.1 Chapter overview

This chapter will focus on how to structure the random effects for a growth curve analysis. The general principle is *keep it maximal* (Barr et al., 2013): the random effects should include as much of the structure of the data as possible. With this principle in mind, we will extend the between-participants GCA described in the previous chapter to deal with within-participant effects.

These practical issues will then lead to a more general discussion of whether participants should be treated as fixed or random effects. In addition to answering this question, this discussion will provide a deeper understanding of the difference between fixed and random effects, which is critical to understanding multilevel modeling and will be of particular relevance to researchers interested in individual differences.

The final section will describe how to use visual comparisons to aid in the interpretation of polynomial time terms. At first blush, this may not seem to be related to structuring random effects, but, as we will see, this kind of visual comparison requires manipulating random effects.

4.2   "Keep it maximal"

Recall from Chapter 2 that random effects correspond to the observational units in the study and capture the nested structure of the data. A *full* or *maximal* random effect structure is the case where all of the factors that could hypothetically vary across individual observational units are allowed to do so. Let's revisit the word learning model from the previous chapter:

```
> m.2 <- lmer(Accuracy ~ (ot1+ot2)*TP + (ot1+ot2 | Subject),
              data=WordLearnEx, REML=FALSE)
```

The Level 1 model describes the overall word learning time course as a second-order orthogonal polynomial. The observational units — the study participants — are assumed to represent random deviations from this overall pattern. This means that, in principle, individual participants could vary on any of the three time terms: the intercept, linear, or quadratic terms. To capture this, we include all of them in the random effects specification (recall that the intercept is included by default):

```
> (ot1+ot2 | Subject)
```

For this model, this is the *full* or *maximal* random effect structure. Table 4.1 contains the parameter estimates for this model. What would happen if we

TABLE 4.1

Parameter Estimates Using Maximal Random Effect Structure

	Estimate	Std. Error	t	p
Intercept	0.779	0.022	35.831	0.000
Linear	0.286	0.038	7.577	0.000
Quadratic	-0.051	0.033	-1.532	0.125
High TP: Intercept	0.053	0.031	1.724	0.085
High TP: Linear	0.001	0.053	0.020	0.984
High TP: Quadratic	-0.116	0.047	-2.481	0.013

omitted one of the time terms from the random effects structure? Here is the same model but with only the intercept and linear terms in the random effects:

```
> m.1r <- lmer(Accuracy ~ (ot1+ot2)*TP + (ot1 | Subject),
               data=WordLearnEx, REML=FALSE)
```

The new parameter estimates (Table 4.2) are exactly the same, but the standard errors are smaller, which makes their effects more statistically significant. Omitting the quadratic random effect term told the model that all participants should have the same quadratic term, which pushed all of the variability to the fixed effects, thus making them look more significant. More generally, Barr

TABLE 4.2

Parameter Estimates Using only Intercept and Linear Terms in the Random Effect Structure

	Estimate	Std. Error	t	p
Intercept	0.779	0.022	35.847	0.000
Linear	0.286	0.037	7.789	0.000
Quadratic	-0.051	0.030	-1.681	0.093
High TP: Intercept	0.053	0.031	1.724	0.085
High TP: Linear	0.001	0.052	0.021	0.983
High TP: Quadratic	-0.116	0.043	-2.722	0.006

et al. (2013) used Monte Carlo simulation to show that maximal random effects structures minimize false alarm rates without substantial loss of power (for additional discussion of how to structure random effects for mixed designs with both within-subject and between-subject factors, see Barr, 2013).

As described in Chapter 2, growth curve models are fit using an iterative algorithm that tries to find (*converge to*) the set of parameters that maximizes the likelihood of observing the actual data. Models with complex random effects structures can take a long time to converge and can fail to converge. If the maximal model does not converge, it may be necessary to simplify the random effect structure, that is, to remove some terms from the random effects. As we saw in the word learning example, removing a time term from the random effects primarily reduces the standard error of the corresponding fixed effect estimate, making it look more significant and (based on the results of Barr et al., 2013) inflating the false alarm rate. This means that the most important random effect terms are those that correspond to the fixed effects of interest. For example, in higher-order polynomial models, some of the higher-order terms may be of less interest and thus good candidates for removal from random effects. But remember that this is only a last resort if a model with the maximal random effect structure fails to converge.

Another approach is to remove the random effect correlations. By default, the random effects structure specified by

```
> (ot1+ot2 | Subject)
```

tells `lmer` to estimate the three participant-level random effects (intercept, linear, and quadratic) and their pairwise correlations. You can see these correlations on the right side of the **Random effects** section near the beginning of the model summary, excerpted here:

```
Groups    Name        Std.Dev. Corr
Subject   (Intercept) 0.1037
          ot1         0.1242   -0.33
          ot2         0.0792   -0.28 -0.82
Residual              0.1567
```

It is possible to remove those correlations, effectively setting them to 0, by specifying the three random effects separately using the following syntax:

```
> m.nocorr <- lmer(Accuracy ~ (ot1+ot2)*TP + (1 | Subject) +
                   (0+ot1 | Subject) + (0+ot2 | Subject),
                   data=WordLearnEx, REML=FALSE)
```

Now those correlations are absent from the summary of the random effects (notice that the residual variance is now larger because these reduced random effects capture less variance):

```
Groups      Name        Std.Dev.
Subject     (Intercept) 0.1032
Subject.1   ot1         0.1105
Subject.2   ot2         0.0105
Residual                0.1600
```

As with removing other random effect terms, this does not affect the fixed effect parameter estimates, but it does affect their standard errors (see Table 4.3). In their simulations, Barr et al. (2013) found that removing random effect

TABLE 4.3
Parameter Estimates for Model without Random Effect Correlations

	Estimate	Std. Error	t	p
Intercept	0.779	0.022	35.847	0.000
Linear	0.286	0.037	7.789	0.000
Quadratic	-0.051	0.030	-1.678	0.093
High TP: Intercept	0.053	0.031	1.724	0.085
High TP: Linear	0.001	0.052	0.021	0.983
High TP: Quadratic	-0.116	0.043	-2.717	0.007

correlations resulted in the smallest increase in false alarm rates, so these may also be good candidates for removal when it is necessary to simplify the random effect structure. Keep in mind, however, that Barr et al. did not examine time course data and these correlations may play a bigger role for time course data. For example, note the very strong negative correlation between ot1 and ot2 random effects, which probably reflects the fact that individuals who learned the words faster have steeper positive slopes (higher ot1) and more sharply (negatively) curved learning curves (lower ot2) – Chapter 7 will discuss how to use random effects to quantify individual differences in more detail. Until systematic analyses of the sort conducted by Barr et al. are conducted for time course data, it remains unclear how much the false alarm rate will be inflated by removing the random effect correlations in analyses of time course data.

To summarize, the best strategy is to start with a maximal random effects structure that includes all time terms. If this model fails to converge, it may help to simplify the random effects structure by removing some of the terms. Removing terms from the random effects will not change the fixed effect estimates, but it will change their standard errors, which will make them look more statistically significant and possibly increase the likelihood of a Type I error (i.e., a false positive). If you need to simplify the random effect structure, the best candidates appear to be the random effect correlations and random effects corresponding to fixed effects that are not of primary interest.

4.3 Within-participant effects

So far, all of the examples have considered only between-participant manipulations where there is only one time series of observations per participant. In many studies there are multiple time series per participant, for example, due to within-participant manipulations where each participant completes multiple conditions. In such cases this additional level of nesting needs to be represented in the random effects. Let's walk through an example using data on the time course of word recognition. In this study, participants were shown four pictures on a computer screen, then they heard the name of one of the pictures and had to click on it. There were two kinds of words: words like *horse* and *bed*, which occur frequently in typical language use ("high" frequency words) and words like *mouse* and *comb*, which occur less frequently in typical language use ("low" frequency words). Each participant heard both kinds of words, so this was a within-participant manipulation. The participants' eye movements were tracked as they completed this word-to-picture matching task and the data correspond to each participant's probability of fixating the target picture at each point in time.

```
> summary(TargetFix)
    Subject         Time          timeBin      Condition
 708    : 30   Min.   : 300   Min.   : 1   High:150
 712    : 30   1st Qu.: 450   1st Qu.: 4   Low :150
 715    : 30   Median : 650   Median : 8
 720    : 30   Mean   : 650   Mean   : 8
 722    : 30   3rd Qu.: 850   3rd Qu.:12
 725    : 30   Max.   :1000   Max.   :15
 (Other):120
    meanFix            sumFix            N
 Min.   :0.0286   Min.   : 1.0   Min.   :33.0
 1st Qu.:0.2778   1st Qu.:10.0   1st Qu.:35.8
 Median :0.4558   Median :16.0   Median :36.0
 Mean   :0.4483   Mean   :15.9   Mean   :35.5
 3rd Qu.:0.6111   3rd Qu.:21.2   3rd Qu.:36.0
 Max.   :0.8286   Max.   :29.0   Max.   :36.0
```

The `TargetFix` data frame contains the following variables:

- `Subject`: a participant ID

- `Time`: milliseconds from the start of the spoken word, ranging from 300ms to 1000ms in 50ms time bins

- `timeBin`: the same time bins represented as 1 to 15 for convenience in fitting the polynomials

- `Condition`: the word type, high frequency or low frequency

- `meanFix`: the proportion of trials on which the target picture was fixated for each participant, in each condition, in each time bin

- `sumFix` and `N`: the numerator and denominator, respectively, for computing the fixation proportions. These will be relevant in Chapter 6, when we discuss growth curve analyses for binary outcome variables (i.e., looking vs. not-looking at the target)

Here is code to plot these data, using a semi-transparent `ribbon` to represent the standard error (Figure 4.1). The `color=NA` option removes lines from the ribbon edges and the `alpha=0.3` sets the level of transparency/opacity.

```
> ggplot(TargetFix, aes(Time, meanFix, linetype=Condition)) +
    stat_summary(fun.data=mean_se, geom="ribbon",
                 color=NA, alpha=0.3) +
    stat_summary(fun.y=mean, geom="line") +
    theme_bw(base_size=10) +
    labs(y="Fixation Proportion",
         x="Time since word onset (ms)")
```

The target fixation proportion starts out approximately at chance (there were four pictures, so chance probability is 25%) and gradually rises until most looks are on the target, indicating that participants have recognized the word and found the corresponding picture. It is clear (and not surprising) that high frequency words were recognized faster than low frequency words.

As with the word learning data, we start by creating an orthogonal polynomial. The target fixation data have three changes of direction (inflection points): the initial change from flatness, then an early increase when fixation proportions begin to rise rapidly, and the last near the end when they begin to plateau. Using the general guideline from the previous chapter, this means that we will need a third-order polynomial to capture these data.

```
> t <- poly(unique(TargetFix$timeBin), 3)
```

then we append it to the `TargetFix` data frame, making sure to align the orthogonal polynomial values with their corresponding time bins.

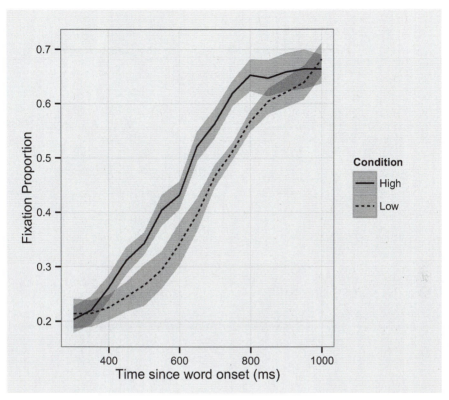

FIGURE 4.1

Target fixation time course for high and low frequency words. Ribbon represents ±SE.

```
> TargetFix[,paste("ot", 1:3, sep="")] <-
                            t[TargetFix$timeBin, 1:3]
```

For convenience, we'll skip to the full model and use the normal approximation to get *p*-values. We'll also use the **bobyqa** optimizer instead of the default **Nelder-Mead** optimizer – in general, the two optimizers produce almost identical results, but **bobyqa** seems to converge somewhat more reliably.

```
> m.full <- lmer(meanFix ~ (ot1+ot2+ot3)*Condition +
                    (ot1+ot2+ot3 | Subject) +
                    (ot1+ot2+ot3 | Subject:Condition),
              control=lmerControl(optimizer = "bobyqa"),
              data=TargetFix, REML=FALSE)
```

The first line contains the fixed effects and the second and third lines contain the model's two sets of random effects. The first set

```
> (ot1+ot2+ot3 | Subject)
```

captures participant-level variability in overall spoken word recognition time course, that is, across both word frequency conditions. The second set

```
> (ot1+ot2+ot3 | Subject:Condition)
```

captures participant-by-condition variability; that is, differences in individual participants' sensitivity to the manipulation. To get a better understanding of what this means, it will help to look at the actual random effect values, which can be extracted from the model object using the **ranef** function:

```
> ranef(m.full)
$`Subject:Condition`
          (Intercept)        ot1         ot2          ot3
708:High    0.0122778 -0.131207  -0.1298541   0.01514974
708:Low    -0.0612169  0.170376   0.0622766   0.01231449
712:High    0.0212282  0.082904   0.0280306   0.02000116
712:Low    -0.0144528  0.044843   0.0325511  -0.01729650
715:High    0.0123634  0.059716   0.0551747  -0.02499504
715:Low    -0.0086839  0.105990   0.1303139  -0.03947449
720:High    0.0100855  0.083385   0.0745981  -0.04445714
720:Low     0.0347270 -0.354064  -0.3136092   0.07583080
722:High    0.0161374 -0.083817  -0.2057153   0.00095489
722:Low    -0.0146185  0.048801   0.0978763  -0.01446828
725:High   -0.0360243  0.019823   0.0763439   0.00073756
725:Low     0.0680030 -0.115198  -0.0433394   0.02575001
726:High    0.0732496 -0.019889   0.0464758   0.00593915
726:Low    -0.0211906  0.031605   0.0195368   0.00968963
730:High   -0.0646790  0.188067   0.0932936   0.02108594
730:Low     0.0132856 -0.018365   0.0416385  -0.03930511
734:High    0.0303379 -0.213237  -0.0452559   0.03463086
734:Low     0.0158956  0.114272  -0.0129925  -0.02218751
736:High   -0.0749766  0.014255   0.0069084  -0.02904712
736:Low    -0.0117485 -0.028259  -0.0142522   0.00914695


$Subject
      (Intercept)        ot1         ot2          ot3
708 -0.00015795  0.0108620  -0.0035599  -0.00493085
712  0.01102197  0.1069207  -0.0093299  -0.03651407
715  0.01137522  0.1148167  -0.0109608  -0.03965103
720 -0.00207073 -0.0130026  -0.0003585   0.00374228
722  0.01382768  0.1622380  -0.0200789  -0.05817441
725 -0.01782716 -0.2073434   0.0253441   0.07419975
726 -0.00020340 -0.0269480   0.0076148   0.01166406
730 -0.00172924 -0.0167750   0.0014638   0.00572876
734  0.00105459  0.0033764   0.0011498  -0.00047767
736 -0.01529097 -0.1341449   0.0087155   0.04441316
```

```
attr(,"class")
[1] "ranef.mer"
```

The `ranef` function returns a list of two data frames, corresponding to the two sets of random effects and named by the term to the right of the pipe in the model specification. Let's start with the second data frame in the list, which contains the `Subject`-level random effects. The rows are the individual participants and the columns are the time terms. The values are the random effects, which capture how that individual's time course differed from the group mean with respect to that particular time term. Looking at the first row, we see that participant #708 had an intercept that was very close to the average (-0.00016), a linear slope that was somewhat more positive than the average (0.0109), a quadratic curvature that was slightly more negative than average (-0.0036), and a cubic curvature that was slightly more negative than the average (-0.0049). These values describe each participant's deviation from the overall group pattern in terms of the time variables that were used to describe the overall group pattern. Notice that condition is not represented in these random effects, so they correspond to overall, condition-independent, individual differences.

Now let's go back to the first data frame in the list, `Subject:Condition`, which contains a random effect estimate for each participant-by-condition combination for each time term. The principle is the same, except now we have moved one level down in the nesting hierarchy to consider separately the two sets of observations for each participant – one set for each of the within-participant conditions. For example, taking the first two values in the left-most column, this says that the intercept for participant #708 in the High condition was slightly higher than the overall average (0.0123) and the intercept in the Low condition was somewhat lower than the overall average (-0.0612). Chapter 7 will describe how these values can be used to calculate effect sizes for individual participants.

Together, the two sets of random effects capture the expected variability in the data at the individual participant level (the Subject random effects) and at the participant-by-condition level (Subject-by-Condition random effects) as well as the nested structure of the data (individual observations grouped by condition and by participant).

4.3.1 An alternative within-participant random effect structure

The focus of this book is on using multilevel regression to analyze time course data, but the same general approach can be applied to other kinds of nested data. In time course data, multiple observations from the same individual are related by a continuous time variable; in other nested data, the multiple observations might simply correspond to different trials on which there was a different (discrete) stimulus item. For example, in our study of recognition of high and low frequency words, there were 36 words of each type. For such data, multilevel regression models are a useful tool for simultaneously capturing the random variability among participants and among items (e.g., Baayen et al., 2008; Barr et al., 2013). In such *crossed* random effects models it is typical to place the condition random effects on the left side of the pipe in the random effect structure, which are sometimes called *random slopes of condition* (this terminology is potentially confusing with time slopes, so we'll continue to discuss them in terms of the left and right side of the pipe). Here is how that approach would look for the target fixation example data:

```
> m.Left <- lmer(meanFix ~ (ot1+ot2+ot3)*Condition +
                          ((ot1+ot2+ot3)*Condition | Subject),
                control=lmerControl(optimizer = "bobyqa"),
                data=TargetFix, REML=FALSE)
```

Looking at the random effect estimates from this model might help with understanding how this is different from the random effects structure described in the previous section.

```
> ranef(m.Left)
$Subject
      (Intercept)        ot1        ot2        ot3 ConditionLow
708    0.0129823 -0.127326 -0.1404948 -0.0213955    -0.071949
712    0.0355779  0.192963  0.0173607 -0.0041654    -0.039625
715    0.0261574  0.182484  0.0027405 -0.0016630    -0.026146
720    0.0091999  0.064358  0.0806459 -0.0635344     0.022465
722    0.0280798  0.087190 -0.2205424 -0.0824442    -0.034123
725   -0.0579212 -0.198721  0.1311462  0.0682712     0.106360
726    0.0722703 -0.042654  0.0651959  0.0257667    -0.088813
730   -0.0665132  0.176032  0.0988989  0.0374008     0.079989
734    0.0335393 -0.210460 -0.0313207  0.0383864    -0.018863
736   -0.0933725 -0.123867 -0.0036302  0.0033772     0.070706
      ot1:ConditionLow ot2:ConditionLow ot3:ConditionLow
708           0.315895        0.2030291        0.0438445
712          -0.054206        0.0228135       -0.0767239
715           0.015228        0.1157983       -0.1061287
720          -0.441935       -0.3913959        0.1671132
722           0.135545        0.2958944        0.0073066
725          -0.111273       -0.1766832        0.0240645
726           0.082522       -0.0703233        0.0154962
730          -0.226211       -0.0320487       -0.1061425
734           0.323745        0.0363369       -0.0378409
736          -0.039309       -0.0034211        0.0690110

attr(,"class")
[1] "ranef.mer"
```

For each participant, there are 8 random effect estimates corresponding to each of the two conditions (High and Low) by each of the four time terms (intercept, slope, quadratic, and cubic). This is different from the random effect structure in the previous section in two important ways.

First, since random effects are meant to model the variability in the data in terms of random samples from a normal distribution, it represents a different idea about how that sampling works. Putting `Condition` on the right side of the pipe treats every participant-by-condition time series as an individual sample drawn from a single distribution, which means estimating two variance parameters per time term (one for `Subject` and one for `Subject:Condition`) and two correlation parameters for each pair of time terms. These estimates are printed near the top of the model `summary`:

```
Groups              Name         Std.Dev. Corr
Subject:Condition  (Intercept)  0.0405
                    ot1          0.1404   -0.43
                    ot2          0.1124   -0.33  0.72
                    ot3          0.0417    0.13 -0.49 -0.43
Subject            (Intercept)  0.0124
                    ot1          0.1195    0.91
                    ot2          0.0165   -0.42 -0.76
                    ot3          0.0421   -0.85 -0.99  0.83
Residual                         0.0438
```

In contrast, putting `Condition` on the left side creates separate distributions for each condition, which means estimating separate variance parameters per level of `Condition` for each time term and separate correlation parameters for each time term-by-level combination:

```
Groups    Name             Std.Dev. Corr
Subject   (Intercept)      0.0519
          ot1              0.1570    0.18
          ot2              0.1094   -0.29  0.03
          ot3              0.0490   -0.28 -0.37  0.63
          ConditionLow     0.0649   -0.89 -0.13  0.49  0.34
          ot1:ConditionLow 0.2285    0.38 -0.46 -0.62  0.10 -0.56
          ot2:ConditionLow 0.1889    0.20  0.08 -0.81 -0.22 -0.43  0.74
          ot3:ConditionLow 0.0862   -0.08 -0.40 -0.06 -0.47  0.06 -0.27 -0.43
Residual                   0.0430
```

The upshot is that putting `Condition` on the left side creates a more flexible model with fewer assumptions (e.g., not assuming equal variance across conditions), but at the cost of a substantial increase in the number of parameters that need to be estimated. For this example, with 4 time terms and 2 conditions, it is a difference between 36 and 20 parameters. This difference in number of parameters also means that models with `Condition` on the left side of the pipe will take longer to fit (for this example it was 0.92sec vs. 2.84sec) and are more likely to fail to converge.

Second, having separate participant and participant-by-condition random effects allows for participant-level random effect estimates that are independent of individual conditions (that is, collapsed across the two conditions). Such random effects might be useful for describing individual participants' overall time course. For example, participant #722 had a very positive ot1 random effect estimate (0.162) and a very negative ot2 random effect estimate (-0.02), suggesting that his or her overall word recognition was much faster than average. The alternative structure with Condition to the left of the pipe does not provide this kind of estimate, though an analog could be computed by averaging together the two condition-specific random effect estimates for each participant.

```
> 0.5 *
    (ranef(m.Left)$Subject[,1:4] + ranef(m.Left)$Subject[,5:8])
       (Intercept)        ot1         ot2           ot3
708 -2.9483e-02   0.094284   0.0312672    0.01122449
712 -2.0238e-03   0.069379   0.0200871   -0.04044464
715  5.7761e-06   0.098856   0.0592694   -0.05389581
720  1.5832e-02  -0.188788  -0.1553750    0.05178941
722 -3.0215e-03   0.111368   0.0376760   -0.03756879
725  2.4219e-02  -0.154997  -0.0227685    0.04616784
726 -8.2715e-03   0.019934  -0.0025637    0.02063144
730  6.7378e-03  -0.025090   0.0334251   -0.03437083
734  7.3381e-03   0.056642   0.0025081    0.00027279
736 -1.1333e-02  -0.081588  -0.0035257    0.03619410
```

The consequences of these differences are not obvious without a thorough and systematic analysis, but the additional assumptions are relatively reasonable (e.g., equal variance across conditions) and the benefits in terms of computational time and convergence are noticeable. For these reasons, specifying separate Subject and Subject-by-Condition random effects as in section 4.3 is the recommended approach, but the alternative can be used if a more flexible model is needed.

4.4 Participants as random vs. fixed effects

The conceptual overview in Chapter 2 described the traditional logic that if a factor is interesting in itself and its levels are fixed in the world and reproducible, then they should be considered fixed effects; if the levels correspond to randomly sampled observational units, then they should be considered random effects. Following this logic, the preceding examples all treated participants as random effects. However, researchers interpret this logic in different ways. For example, in two articles describing how to apply multilevel regression to fixation time course data, one treated participants as random effects (Barr, 2008) and the other treated them as fixed effects (Mirman, Dixon, & Magnuson, 2008). This section will work through the consequences of this decision and provide some guidelines for choosing which approach is more appropriate for a particular situation. In general, this discussion will focus on participant effects because our primary analyses typically focus on data aggregated by participants, but the same issues and principles hold for items in a by-items analysis.

The critical difference is that when participants are treated as a fixed effect, each participant's parameters are estimated independently. When participants are treated as a random effect, each participant's parameters are constrained to be random deviations from the population mean parameters, with the deviations assumed to conform to a normal distribution with mean equal to 0. This additional constraint means that each individual's parameter estimates from a random-participant-effects model will be weighted averages of the parameter estimates from a fixed-participant-effects model and the group-level parameter estimates. Put simply, the parameter estimates reflect both the individual participant's data and the whole group data – each participant's individual random effect parameter estimates are influenced by the other participants' data. As a result, they tend to "shrink" toward the population mean. This shrinkage can have positive and negative consequences. When individual participant estimates are allowed to be fully independent (i.e., treated as fixed effects), they provide better (that is, independent) estimates of differences between individual participants. The downside is that the resulting model can overfit the data – modeling every tiny difference between participants may tell us more about the trees than the forest.

Figure 4.2 shows an example of this shrinkage. The data were taken from a study that used eye tracking to examine semantic competition effects during spoken word comprehension (Mirman & Magnuson, 2009; the same data were used to demonstrate dynamic inconsistency in Chapter 3). While doing a spoken word-to-picture matching task, participants were (briefly) more likely to look at pictures that were semantically related to the target than unrelated pictures. The overall time course of this effect (Level 1) was modeled with a fourth-order orthogonal polynomial, with Level 2 fixed effects of picture relatedness (related vs. unrelated), and participants treated as either fixed or random effects. Figure 4.2 shows the effect of this difference on the intercept and linear parameter estimates for individual participants. When participants were treated as random effects, the estimates were much more tightly clustered around the population mean (indicated by the black vertical and horizontal lines).

It is important to keep in mind that although this decision affected the individual participant parameter estimates, the condition parameter estimates (related vs. unrelated) came out exactly the same because this study had a balanced within-participant design (i.e., there were related and unrelated distractors for all participants). However, because of the additional constraints

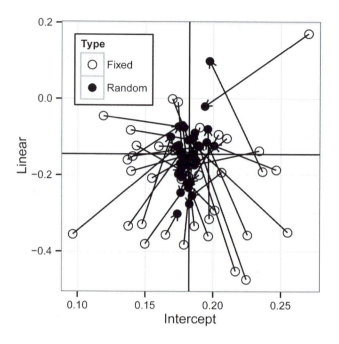

FIGURE 4.2

Shrinkage effect on individual participant intercept and linear term parameter estimates. For each participant, the arrow shows the change in the parameter estimate from a model that treats participants as fixed effects (open circles) to a model that treats participants as random effects (filled circles). The black vertical and horizontal lines indicate the population-level fixed effect.

of treating participants as random effects, the participant effects captured less variance, thus the standard errors for the condition fixed effect parameter estimates were larger. As a result, treating participants as random effects yields a more conservative estimate of group-level effects.

Another important consequence is that treating participants as fixed effects gives the model much more flexibility, so it is able to attain a much better model fit (participants as fixed effect: LL = 1189; participants as random effect: LL = 1024). The log-likelihood always increases when independent parameters are added to a model, but, as described in Chapter 2, we can test whether the additional parameters significantly improve model fit by evaluating the change in the deviance statistic $(-2 \cdot \Delta LL)$. In this case, the additional participant fixed effect parameters did significantly improve model fit $(\chi^2(170) = 329, p < 0.0001)$. This example illustrates a fundamental tension between building a model that provides the best statistical description of the data and a model that instantiates the research question. In many studies (including the study from which these data were drawn), researchers are interested in generalizing from their sample to a larger population, so the individual variability reflects the variability in the population. Treating participants as random effects instantiates this premise in a way that makes the condition fixed effect estimates more conservative, so it is appropriate for that research question.

When deciding whether to treat participants as random or fixed effects, one needs to consider the research goals and the homogeneity and normality of the sample population. If the goal is to generalize, then the researcher is essentially forced to assume that the sample is drawn from a homogeneous population (otherwise generalization would be impossible) and should treat participants as a random effect. However, this form of generalization is not always the goal. For example, neurological case studies inform cognitive theories by showing what must be possible (as in an existence proof) and by generating new hypotheses. In such contexts, the goal is to describe the observed data as accurately as possible and treating participants as fixed effects may be more appropriate. Since participant fixed effect parameters better capture individual differences, they may provide a better approach for studying individual differences (e.g., Mirman et al., 2011; and the individual differences example in Mirman, Dixon, & Magnuson, 2008). In such cases, it may be advantageous to acquire independent parameter estimates for the participants by treating them as fixed effects rather than random effects. Finally, for hypothetically homogeneous populations like typical college students, treating participants as random effects may the better approach; but for clearly non-homogeneous populations like neurological patients (who have unique clinical and neurological presentations, even if their diagnosis is the same) treating participants as fixed effects may be more appropriate.

In sum, for typical experiments, treating participants (or items) as random effects appropriately reflects the typical assumption that each observational unit is a randomly drawn sample from the population to which the researcher hopes to generalize. Treating participants as fixed effects is a legitimate alternative, but should be explicitly justified based on sample properties (e.g., non-random sampling from a non-homogeneous or non-normal distribution) or research goals (e.g., description of present data rather than generalization to a population).

4.5 Visualizing effects of polynomial time terms

One of the challenges of using polynomial functions in growth curve analyses is interpreting effects on polynomial time terms, especially the higher-order terms. Interpreting the effects of lower-order terms like the intercept and linear slope is relatively easy, but it can be hard to know what aspect of the time course is being captured by a significant effect on the cubic or quartic term.

One solution to this problem is to just not include effects on these higher-order terms in the model. However, although they can be difficult to interpret, the higher-order terms can capture interesting and important aspects of the data. In such cases, direct visual comparisons can help both the researcher and the reader understand those effects. We'll work through an example where the cubic and quartic terms capture an important effect and demonstrate how to use visual comparisons to interpret those results.

The example data come from an experiment using the same eye-tracking paradigm as the other examples in this chapter. On each trial in this experiment, participants had to click on the picture that matched a spoken word. The display contained four pictures: the target and three distractors. Two of those distractors were completely unrelated to the target, but one distractor was related to the target either because they had a similar general function (for example, *toaster* and *coffee-maker* are both used to prepare breakfast; this was called the *function* condition) or because they were typically used together in some way (for example, *toaster* and *bread* are used together to make toast; this was called the *thematic* condition). The participants tended to look at these related distractors (*competitors*) more than at the unrelated distractors and, more importantly, the looks to the thematic competitor tended to happen earlier than the looks to the function competitor, suggesting that this kind of relation was recognized more quickly (Kalénine et al., 2012). Here is a data frame containing the relevant data:

```
> summary(FunctTheme)
    Subject          Time          meanFix           Condition
 21     :102   Min.   : 500   Min.   :0.0000   Function:765
 24     :102   1st Qu.: 700   1st Qu.:0.0625   Thematic:765
 25     :102   Median : 900   Median :0.1333
 27     :102   Mean   : 900   Mean   :0.2278
 28     :102   3rd Qu.:1100   3rd Qu.:0.3113
 40     :102   Max.   :1300   Max.   :1.0000
 (Other):918
          Object
 Target     :510
 Competitor:510
 Unrelated :510
```

Here is a plot of those data (Figure 4.3):

```
> ggplot(FunctTheme, aes(Time, meanFix, linetype=Object)) +
    facet_wrap(~ Condition) +
    stat_summary(fun.y=mean, geom="line") +
    stat_summary(fun.data=mean_se, geom="ribbon",
                 color=NA, alpha=0.3) +
    theme_bw(base_size=10) +
    labs(x="Time Since Word Onset (ms)",
```

```
            y="Fixation Proportion") +
     theme(legend.justification=c(0,1),
           legend.position=c(0,1),
           legend.background=
              element_rect(color="black", fill="white")) +
     scale_linetype_manual(values=
                          c("solid", "dashed", "dotted"))
```

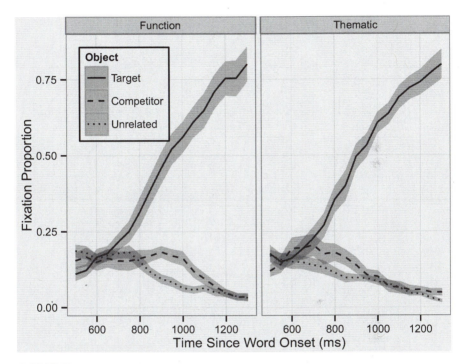

FIGURE 4.3
Time course of fixations to targets (solid lines), related competitors (dashed lines), and unrelated distractors (dotted lines) in the function (left panel) and thematic (right panel) relation conditions. Ribbon represents ±SE.

To prepare the data for growth curve analysis, it will be convenient to have a `timeBin` variable that represents time as bins in the analysis instead of milliseconds since word onset

```
> FunctTheme$timeBin <- FunctTheme$Time/50 - 9
```

Now we can create a fourth-order orthogonal polynomial in the range of `time-Bin`

```
> t <- poly(1:max(FunctTheme$timeBin), 4)
```

insert it into the data frame, aligned by `timeBin`

```
> FunctTheme[, paste("ot", 1:4, sep="")] <-
                              t[FunctTheme$timeBin, 1:4]
```

and fit the full model:

```
> m.full <- lmer(meanFix ~ (ot1+ot2+ot3+ot4)*Object*Condition +
                 (ot1+ot2+ot3+ot4 | Subject) +
                 (ot1+ot2+ot3+ot4 | Subject:Object:Condition),
                 data=subset(FunctTheme, Object != "Target"),
                 control=lmerControl(optimizer="bobyqa"),
                 REML=FALSE)
```

The fixed effects contain all four polynomial time terms (and the inter-cept is included by default) and all of their interactions with `Object` and with `Condition` and the `Object:Condition` interaction. The effects of `Object` will capture the difference between related and unrelated distractors; the effects of `Condition` will capture the differences between the Function and Thematic conditions; and the `Object:Condition` interaction is the most important be-cause these effects are the ones that will capture how the type of relation (Condition) modulates the time course of looks to the related vs. unrelated distractors. As in section 4.3, there are two sets of random effects: one set at the `Subject` level to capture overall individual differences and one set at the lowest level of nesting in the data (`Subject:Object:Condition`). Finally, because we are only interested in comparing the related vs. unrelated distrac-tors, we exclude the target fixations from the analysis data using the `subset` function.

We can plot the model fit to check that the model actually fit the data reasonably well. For convenience, we'll combine the subset of behavioral data that was analyzed (exclude the target) and the model fit into one data frame:

```
> data.comp <- data.frame(
                  subset(FunctTheme, Object != "Target"),
                  GCA_Full=fitted(m.full))
```

This makes it easier to make a combined plot of the behavioral data and model fit (Figure 4.4):

```
> ggplot(data.comp, aes(Time, meanFix, shape=Object)) +
    facet_wrap(~ Condition) +
    stat_summary(fun.data=mean_se, geom="pointrange") +
    stat_summary(aes(y=GCA_Full, linetype=Object),
                 fun.y=mean, geom="line") +
    theme_bw(base_size=10) +
    labs(x="Time Since Word Onset (ms)",
        y="Fixation Proportion") +
    theme(legend.justification=c(1,1),
        legend.position=c(1,1),
```

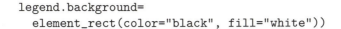

```
legend.background=
  element_rect(color="black", fill="white"))
```

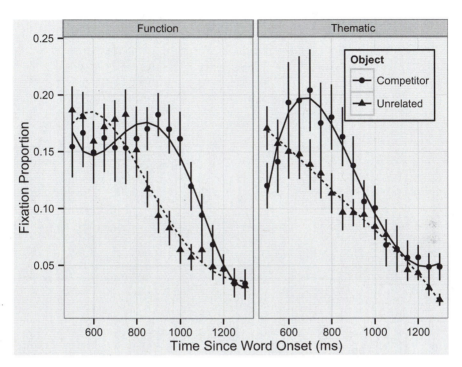

FIGURE 4.4

Time course of fixations to related competitors (circles, solid lines) and unrelated distractors (triangles, dashed lines) in the function (left panel) and thematic (right panel) relation conditions. Symbols represent behavioral data (±SE); lines represent full GCA model fits.

The model fits the behavioral data quite well. To evaluate the experimental effects, we can compute the p-values using the normal approximation:

```
> coefs.full <- as.data.frame(coef(summary(m.full)))
> coefs.full$p <- format.pval(
                2*(1-pnorm(abs(coefs.full[,"t value"]))))
```

and extract the critical `Object:Condition` interaction parameters using the `str_detect` function from the `stringr` package:

```
> library(stringr)
> coefs.full[str_detect(rownames(coefs.full),
          "*ObjectUnrelated:ConditionThematic"),]
                                  Estimate Std. Error
ObjectUnrelated:ConditionThematic -0.0041636  0.017089
```

```
ot1:ObjectUnrelated:ConditionThematic   0.0658778   0.077445
ot2:ObjectUnrelated:ConditionThematic  -0.0475679   0.043617
ot3:ObjectUnrelated:ConditionThematic  -0.1561839   0.051811
ot4:ObjectUnrelated:ConditionThematic   0.0757088   0.033078
                                          t value          p
ObjectUnrelated:ConditionThematic        -0.24364   0.807512
ot1:ObjectUnrelated:ConditionThematic     0.85064   0.394969
ot2:ObjectUnrelated:ConditionThematic    -1.09059   0.275451
ot3:ObjectUnrelated:ConditionThematic    -3.01450   0.002574
ot4:ObjectUnrelated:ConditionThematic     2.28881   0.022090
```

The statistically significant effects of the interaction are only on the cubic and quartic terms. It would be nice if this statistical difference corresponded to the earlier vs. later competition effect shown in Figure 4.4, but it is hard to be sure because those higher-order terms are hard to mentally visualize. To simplify that, we can physically visualize the effects of those terms by fitting a reduced model with those specific effects removed and then visually comparing the model fits.

For statistical model comparisons, as discussed in Chapter 2, we would only remove the fixed effects and compare the models using the **anova** function. For this sort of visual comparison, we need to remove both the fixed effects and the corresponding random effects. If we just removed the fixed effects, the random effects would pick up some of that variance and it would be hard to see the differences in the plots. Because we will be removing specific interaction terms, we'll have to use a somewhat less compact model formula:

```
> m.red <- lmer(meanFix ~ (ot1+ot2+ot3+ot4)*Object +
                          (ot1+ot2+ot3+ot4)*Condition +
                          (ot1+ot2)*Object*Condition +
                          (ot1+ot2+ot3+ot4 | Subject) +
                          (ot1+ot2 | Subject:Object:Condition),
               control=lmerControl(optimizer="bobyqa"),
               data=subset(FunctTheme, Object != "Target"),
               REML=FALSE)
```

The reduced fixed effect interaction term is in the third line and the reduced random effect interaction term is in the fifth line. For convenience, we'll add this model fit to the data frame that already has the behavioral data and the full model fit.

```
> data.comp$GCA_Reduced <- fitted(m.red)
```

We could try to plot these behavioral data and both model fits, but that would mean plotting 12 time series in one figure, which would be hard to make sense of. Since what we really care about is the difference between the fixation time courses for competitor and the unrelated objects, we can simplify the figure by plotting just the difference between them, which is typically called the "competition effect." First, we have to calculate those differences for each participant, at each time point, in each condition.

The `plyr` package provides tools to do this sort of calculation efficiently. The `plyr` package implements a `split-apply-combine` strategy: the data set is split into subsets, some operation is applied to those subsets, then the results are combined to create a new data set. The key functions in the `plyr` package have names that look like ****ply** where each * corresponds to the format of the input or output data set (respectively) and can be `a` for an array, `l` for a list, or `d` for a data frame. Since we're starting with a data frame and we want to end up with a data frame, we'll use `ddply`:

```
> ES <- ddply(data.comp, .(Subject, Time, Condition),
            summarize,
            Competition = meanFix[Object=="Competitor"] -
                          meanFix[Object=="Unrelated"],
            GCA_Full = GCA_Full[Object=="Competitor"] -
                       GCA_Full[Object=="Unrelated"],
            GCA_Reduced = GCA_Reduced[Object=="Competitor"] -
                          GCA_Reduced[Object=="Unrelated"])
```

The first input to `ddply` is the starting data frame, in our case it is `data.comp`. The next input is a list of variables that define the subsets – there will be a subset for every unique combination of values among these variables. Since we wanted to compute the effect size for each participant, at each time point, in each condition, the list of variables is `.(Subject, Time, Condition)`. After these are specified, we need to tell `ddply` what operation to apply to each subset. In our case, we define a data summary that consists of differences between competitor and unrelated fixations (observed or model-predicted). Here is a summary of the resulting data frame:

```
> summary(ES)
    Subject          Time         Condition      Competition
 21     : 34   Min.   : 500   Function:255   Min.   :-0.2812
 24     : 34   1st Qu.: 700   Thematic:255   1st Qu.:-0.0625
 25     : 34   Median : 900                  Median : 0.0000
 27     : 34   Mean   : 900                  Mean   : 0.0214
 28     : 34   3rd Qu.:1100                  3rd Qu.: 0.0702
 40     : 34   Max.   :1300                  Max.   : 0.4000
 (Other):306
    GCA_Full         GCA_Reduced
 Min.   :-0.2394   Min.   :-0.3123
 1st Qu.:-0.0375   1st Qu.:-0.0231
 Median : 0.0143   Median : 0.0209
 Mean   : 0.0214   Mean   : 0.0215
 3rd Qu.: 0.0760   3rd Qu.: 0.0630
 Max.   : 0.3320   Max.   : 0.2610
```

Now we can plot the observed competition effects in both conditions along with the predictions from the full and reduced models (Figure 4.5).

```
> ggplot(ES, aes(Time, Competition,
                 shape=Condition, linetype=Condition)) +
    stat_summary(fun.y=mean, geom="point") +
    stat_summary(aes(y=GCA_Full), fun.y=mean, geom="line") +
    stat_summary(aes(y=GCA_Reduced), fun.y=mean, geom="line",
                 color="gray") +
    theme_bw(base_size=10) +
    labs(x="Time Since Word Onset (ms)", y="Competition") +
    theme(legend.justification=c(1,1), legend.position=c(1,1),
          legend.background=
                 element_rect(color="black", fill="white")) +
    scale_shape_manual(values=c(1,16))
```

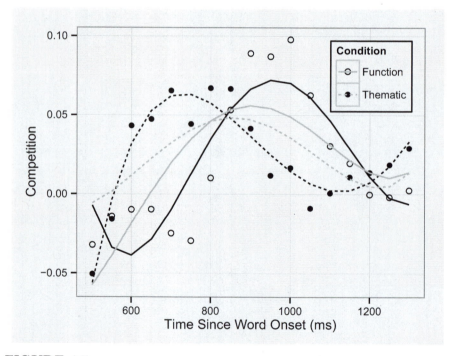

FIGURE 4.5

Time course of competition. Symbols represent the behavioral data, black lines represent the full model, grey lines represent the reduced model.

The full model (black lines) very clearly captured the time course difference between the earlier competition effect for the thematic condition (dashed line) and the later competition effect for the function condition (solid line). After we removed the cubic and quartic terms, the reduced model (grey lines) predicted almost identical time courses for these two conditions. This visual comparison indicates that those statistically significant effects on the higher-order terms really were capturing the early-vs.-late competition difference.

4.6 Chapter recap

This chapter covered several issues regarding structuring random effects for growth curve analysis. The first was the general principle that random effects should be *maximal* – that they should capture as much of the structure of the data as possible. Because maximal random effect structures can cause convergence problems, this chapter also discussed how to simplify random effect structures when necessary. The second major topic was how to structure random effects for within-participant effects. The recommended approach is to use two sets of random effects: one set at the participant level and one at the lowest level in the nested structure, that is, the level of individual time series (e.g., participant-by-condition).

The third topic was whether participants should be treated as random or fixed effects. Treating participants (or items) as random effects captures the standard assumption that they are random samples from the population and the goal is to generalize to that population. However, treating participants as fixed effects can be justified when the assumptions or goals are different. In addition to answering this specific question, discussing the consequences of these two approaches provided an opportunity for a deeper discussion of the difference between fixed and random effects.

The last section in this chapter demonstrated how to use visual comparisons of full and reduced models to interpret the effects of higher-order polynomial time terms. Although the general idea is the same as statistical comparisons, the compared model structures are somewhat different. For statistical model comparisons only one aspect of the model should be different so that the comparison can definitively evaluate the statistical significance of that aspect. Typically, that aspect is a fixed effect and the random effect structure should be kept constant. In contrast, for visual comparisons, both the fixed and corresponding random effects need to be reduced in order to make a clear visual comparison. Note that these two kinds of comparisons are not competing alternatives. They are meant to be used together: statistical comparison for formal evaluation and visual comparison to facilitate interpretation.

4.7   Exercises

1. Use the `TargetFix` data set to explore how random effect structure affects fixed effect estimates. Start from a simple model that only has `Subject` random effects and gradually add `Subject:Condition` random effects (intercept, linear, quadratic, cubic). For the addition of each time term to the `Subject:Condition` random effects, how do each of the fixed effect estimates change? How do their standard errors change? How do their *t*-values and *p*-values change?

2. Make plots showing the effect of *Condition* on each of the time terms (intercept, linear, quadratic, and cubic) for the `TargetFix` data. Use these plots to explain what aspect of the data is captured by each term.

5

Categorical predictors

CONTENTS

5.1 Chapter overview

This chapter will focus on categorical predictor variables. So far, we have primarily discussed how to model the relationship between a continuous predictor, typically *time*, and a continuous outcome, such as *Disability Rating Scale score*. However, like most studies in the psychological and neural sciences, the examples have typically also included a categorical predictor that had discrete levels, such as *control* vs. *treatment* or *high* vs. *low* frequency.

For the purposes of regression, categorical variables need to be converted into numeric values. In most statistical software packages, including R, this happens "under the hood" without requiring the analyst to specify how it should be done, but it is important to understand how that conversion is done because it affects the interpretation of the parameter estimates. Section 5.2 will discuss the two most common ways of doing this: *treatment* or *dummy* coding, which is the default in R, and *sum* or *deviation* coding, which is the most common alternative.

When a categorical predictor has more than two levels or conditions, the analyst may wish to compare different pairs, that is, to evaluate multiple pairwise comparisons. Section 5.3 will cover two approaches to conducting such comparisons: re-fitting the model after re-coding the predictor and using the `multcomp` package to make those comparisons within a single model.

5.2   Coding categorical predictors

5.2.1   In simple linear regression

Coding categorical variables is a general issue that applies to every kind of regression, so let's begin with a very simple linear regression example that does not have any nested observations. Let's say we conducted a survey of 100 people to examine the relationship between income, education level, and gender. Half of the participants were male and half were female, half of each of those groups only completed high school and the other half received a bachelor's degree. Table 5.1 shows our observed means in thousands of dollars (these invented data that are loosely based on U.S. Census Bureau's 2012 Annual Social and Economic Supplement). In a `summary`, the reference level

TABLE 5.1
Mean Income (in $1000) by Education and Gender

	Education	Female	Male	(all)
1	HS	25.962	39.470	32.716
2	College	54.297	84.536	69.417
3	(all)	40.130	62.003	51.067

of a factor is listed first, so we can see that `Female` is the reference level for `Gender` and `HS` is the reference level for `Education`.

```
> summary(dat)
     Subject        Gender        Education        Income
1        : 1    Female:50    HS      :50    Min.    :11.0
2        : 1    Male  :50    College:50    1st Qu.:33.3
3        : 1                               Median :46.3
4        : 1                               Mean    :51.1
5        : 1                               3rd Qu.:69.0
6        : 1                               Max.    :96.5
(Other):94
```

This means that in a simple linear regression, these levels will be treated as the baseline for their respective factors and parameters will be estimated for the other levels:

```
> m <- lm(Income ~ Education*Gender, data=dat)
> summary(m)$coefficients
                   Estimate Std. Error t value
(Intercept)          25.962     1.3616 19.0674
EducationCollege     28.335     1.9256 14.7148
GenderMale           13.508     1.9256  7.0149
```

```
EducationCollege:GenderMale    16.731      2.7232  6.1437
                               Pr(>|t|)
(Intercept)                    2.0688e-34
EducationCollege               2.4094e-26
GenderMale                     3.2337e-10
EducationCollege:GenderMale    1.8255e-08
```

The consequence of this coding scheme is that the (Intercept) parameter corresponds to the mean income level for females with only a high school education, the EducationCollege parameter corresponds to the increase in income for females with a bachelor's degree, the GenderMale parameter corresponds to the income difference between males with a high school education and females with a high school education. In other words, these parameter estimates correspond to the *simple* effects of education or gender at the baseline level of the other predictor. This is called *treatment* or *dummy* coding and is numerically represented by treating the reference or baseline level as 0 and the other level as 1. We can see this by using the contrasts function:

```
> contrasts(dat$Gender)
        Male
Female    0
Male      1
```

Simple effects are informative in some cases, particularly when there is a meaningful baseline or reference level, but often we are interested in estimating the *main* effects of predictors; for example, the average income difference between high school graduates and college graduates across both genders, not just for females. To do this, we need to use *sum* or *deviation* coding. This change can be made using the C function. We'll create a new gender variable so we don't overwrite the original treatment-coded factor:

```
> dat$GenderSum <- C(dat$Gender, sum)
```

The sum-coded factor uses -1 and 1 to represent the two levels, so the baseline (0) is the combination (i.e., sum) of both of the levels.

```
> contrasts(dat$GenderSum)
        [,1]
Female    1
Male     -1
```

When we re-fit the model using the new sum-coded gender factor,

```
> m.sum <- lm(Income ~ Education*GenderSum, data=dat)
```

we see that the parameter estimates now capture the main effect of education:

```
> summary(m.sum)$coefficients
                              Estimate Std. Error t value
(Intercept)                    32.7164    0.96281 33.9803
EducationCollege               36.7004    1.36161 26.9536
GenderSum1                     -6.7540    0.96281 -7.0149
EducationCollege:GenderSum1    -8.3654    1.36161 -6.1437
                              Pr(>|t|)
(Intercept)                   2.5881e-55
EducationCollege              1.4407e-46
GenderSum1                    3.2337e-10
EducationCollege:GenderSum1   1.8255e-08
```

The (Intercept) parameter is now the mean income level for all survey responders with only a high school education (it is equal to the HS row mean in Table 5.1), the EducationCollege parameter is the increase in income for all responders with a bachelor's degree (the difference between the HS and College row means in Table 5.1), and the GenderSum1 parameter is half of the income difference between males with a high school education and females with a high school education (i.e., the difference between the mean for males and the overall mean). In other words, EducationCollege is the main effect of education and GenderSum1 is the simple effect of gender for responders with a high school education. The overall fit of the model is no different, but these parameter estimates may more intuitively answer our research questions. The next section will demonstrate how coding categorical predictors works in growth curve analyses.

5.2.2   In growth curve analysis

Let's revisit the effect of amantadine on recovery from brain injury (first discussed in Chapter 2) and focus on the categorical Group variable.

```
> summary(amant.ex)
    Patient              Group          Week            DRS
   1008   : 5    Placebo    :85    Min.   :0    Min.   : 7.0
   1009   : 5    Amantadine:65    1st Qu.:1    1st Qu.:17.0
   1017   : 5                     Median :2    Median :20.5
   1042   : 5                     Mean   :2    Mean   :19.3
   1044   : 5                     3rd Qu.:3    3rd Qu.:22.0
   1054   : 5                     Max.   :4    Max.   :28.0
   (Other):120
```

The reference level for the Group variable is Placebo (it is listed first in the summary). Just like lm in the previous section, lmer will treat this level as the baseline and estimate parameters for the other level, Amantadine.

```
> m.amant <- lmer(DRS ~ Week*Group + (Week | Patient),
                data=amant.ex, REML=F)
> coef(summary(m.amant))
                       Estimate Std. Error t value
(Intercept)            22.05882    0.48485 45.4964
Week                   -0.70000    0.22117 -3.1650
GroupAmantadine        -1.42805    0.73654 -1.9389
Week:GroupAmantadine   -0.83077    0.33598 -2.4726
```

The first two parameters are the intercept and slope (rate) of recovery specifically for the Placebo group. The next two parameters are the Amantadine group's intercept and slope *relative* to the Placebo group. That is, the Amantadine group started out 1.43 points lower and recovered 0.83 points per week faster than the Placebo group.

For this example, treatment coding is a very sensible approach because the Placebo group is meant to be a baseline and we are interested in whether the Amantadine group differs from this baseline at the start of the study (intercept) and in the rate of recovery (slope). As we saw with the income example, using treatment coding can make the parameter estimates difficult to interpret when there is not an obvious baseline level.

For example, consider data (Figure 5.1) from a motor learning task (like learning to play Guitar Hero) that had a *low* and a *high* difficulty version and 20 participants completed both versions under normal (*control*) conditions and while impaired (such as the influence of alcohol).

```
> summary(MotorLearning)
     SubjID         Difficulty       Condition          Trial
9101    : 120    High:1200    Control :1200    Min.    : 1.0
9103    : 120    Low :1200    Impaired:1200    1st Qu.: 8.0
9105    : 120                                  Median :15.5
9107    : 120                                  Mean   :15.5
9109    : 120                                  3rd Qu.:23.0
9111    : 120                                  Max.   :30.0
(Other):1680
    Accuracy
Min.   :0.000
1st Qu.:0.333
Median :0.750
Mean   :0.634
3rd Qu.:1.000
Max.   :1.000
```

In this case, we have a factorial design with two categorical predictor variables: *difficulty* (low vs. high) and *condition* (control vs. impaired). By default, both of these are coded using treatment contrasts:

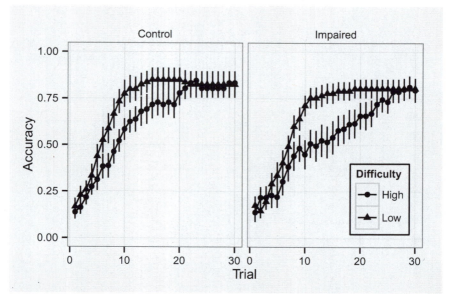

FIGURE 5.1

Accuracy in low and high difficulty versions of a motor learning task under control (left) and impaired (right) conditions. Vertical lines through points indicate ±SE.

```
> contrasts(MotorLearning$Difficulty)
     Low
High  0
Low   1
> contrasts(MotorLearning$Condition)
         Impaired
Control      0
Impaired     1
```

If we fit the model using these defaults, we will see that the parameter estimates will be somewhat counterintuitive and difficult to interpret. First we need to set up a third-order orthogonal polynomial:

```
> t <- poly(1:30, 3)
> MotorLearning[, paste("ot", 1:3, sep="")] <-
                       t[MotorLearning$Trial, 1:3]
```

now we fit the model as usual:

```
> m.ML <- lmer(Accuracy ~ (ot1+ot2+ot3) *
                      Difficulty*Condition +
                      (ot1+ot2+ot3 | SubjID) +
```

```
                    (ot1+ot2+ot3 |
                         SubjID:Difficulty:Condition),
          data=MotorLearning, REML=FALSE)
```

For this model, the base time parameters

```
              Estimate Std. Error t value
(Intercept)   0.621111   0.045521 13.6445
ot1           1.108204   0.135490  8.1792
ot2          -0.441038   0.082074 -5.3737
ot3           0.049899   0.082410  0.6055
```

correspond to the overall baseline level of the data, so these parameters capture the motor learning time course for the high difficulty version in the control condition. The `DifficultyLow` parameter estimates

```
                  Estimate Std. Error t value
DifficultyLow     0.090139   0.049596  1.8175
ot1:DifficultyLow -0.224954   0.159613 -1.4094
ot2:DifficultyLow -0.229082   0.101129 -2.2653
ot3:DifficultyLow  0.184674   0.085386  2.1628
```

correspond to the simple effect of difficulty in the control condition; that is, the time course of motor learning for the low difficulty version in the control condition relative to the high difficulty version in the control condition. The `ConditionImpaired` parameter estimates

```
                     Estimate Std. Error    t value
ConditionImpaired    -0.0798611   0.049596 -1.610223
ot1:ConditionImpaired -0.0469771   0.159613 -0.294319
ot2:ConditionImpaired  0.2823096   0.101129  2.791592
ot3:ConditionImpaired  0.0071204   0.085386  0.083391
```

similarly correspond to the simple effect of condition in the high difficulty version (the high difficulty version in the impaired condition relative to the high difficulty version in the control condition). The final set of parameter estimates

```
                                       Estimate Std. Error
DifficultyLow:ConditionImpaired        0.030417   0.07014
ot1:DifficultyLow:ConditionImpaired    0.143392   0.22573
ot2:DifficultyLow:ConditionImpaired   -0.235602   0.14302
ot3:DifficultyLow:ConditionImpaired   -0.068590   0.12075
                                       t value
DifficultyLow:ConditionImpaired        0.43366
ot1:DifficultyLow:ConditionImpaired    0.63525
ot2:DifficultyLow:ConditionImpaired   -1.64736
ot3:DifficultyLow:ConditionImpaired   -0.56802
```

captures the difference between the low and high difficulty versions in the impaired condition relative to the control condition. The simple effects of difficulty (low vs. high in the control condition) and condition (control vs. impaired for the high difficulty version) are not very informative – it would be more informative if the parameters reflected the overall (main) effects of difficulty (low vs. high across both conditions) and impairment (control vs. impaired across both versions of the task).

There are a few steps we can take to make these parameter estimates easier and more intuitive to interpret. First, the default is for factor levels to be ordered alphabetically with the first level serving as the reference. This happens to be fine for condition because *control* is a sensible reference level and happens to be alphabetically earlier than *impaired*. For difficulty, the *low* difficulty version is a more intuitive baseline than the *high* difficulty version. The **relevel** function can be used to set the reference level for a factor:

```
> MotorLearning$Difficulty <-
                    relevel(MotorLearning$Difficulty, "Low")
```

and we can check that the reference level really did change using **contrasts**:

```
> contrasts(MotorLearning$Difficulty)
     High
Low    0
High   1
```

Second, to have the model estimate main effects instead of simple effects, we need to change the contrasts for both factors from treatment to sum coding. As before, we'll use the C function to change the contrasts and create a new condition variable so we don't overwrite the original treatment-coded factors:

```
> MotorLearning$DifficultySum <-
                    C(MotorLearning$Difficulty, sum)
> MotorLearning$ConditionSum <-
                    C(MotorLearning$Condition, sum)
```

and check the new contrast codings:

```
> contrasts(MotorLearning$DifficultySum)
       [,1]
Low     1
High   -1
> contrasts(MotorLearning$ConditionSum)
           [,1]
Control     1
Impaired   -1
```

When we re-fit the model using these factors, the parameter estimates will be more intuitive:

```
> m.MLsum <- lmer(Accuracy ~ (ot1+ot2+ot3) *
                        DifficultySum*ConditionSum +
                        (ot1+ot2+ot3 | SubjID) +
                        (ot1+ot2+ot3 |
                            SubjID:DifficultySum:ConditionSum),
                data=MotorLearning, REML=FALSE)
```

The base time parameters now correspond to the overall time course of motor learning, which is a sensible baseline:

	Estimate	Std. Error	t value
(Intercept)	0.63385	0.033908	18.6934
ot1	1.00809	0.093829	10.7439
ot2	-0.47332	0.053861	-8.7878
ot3	0.12865	0.063698	2.0197

The `DifficultySum1` parameter estimates

	Estimate	Std. Error	t value
DifficultySum1	0.052674	0.017535	3.0039
ot1:DifficultySum1	-0.076629	0.056432	-1.3579
ot2:DifficultySum1	-0.173441	0.035754	-4.8509
ot3:DifficultySum1	0.075190	0.030188	2.4907

correspond to the overall (main) effect of difficulty; that is, the time course of learning in the low difficulty version (coded as 1) relative to the high difficulty version (coded as -1), across the two conditions. Notice that the estimated effects of difficulty are now substantially stronger because instead of just estimating the simple effect for the control condition, now the main effect is estimated including the impaired condition, where the effect of difficulty was larger.

The `ConditionSum1` parameter estimates

	Estimate	Std. Error	t value
ConditionSum1	0.032326	0.017535	1.84354
ot1:ConditionSum1	-0.012360	0.056432	-0.21902
ot2:ConditionSum1	-0.082254	0.035754	-2.30054
ot3:ConditionSum1	0.013587	0.030188	0.45009

correspond to the main effect of condition: the difference in motor learning time course for the control condition (coded as 1) relative to the impaired condition (coded as -1), across both versions of the task. Notice that the signs on the parameter estimates are reversed compared to the treatment coding because now the control condition is coded as 1 and the impaired condition is coded as -1 instead of 0 and 1, respectively.

The parameter estimates for the highest-level interactions

	Estimate	Std. Error	t value
DifficultySum1:ConditionSum1	-0.0076042	0.017535	-0.43366
ot1:DifficultySum1:ConditionSum1	-0.0358481	0.056432	-0.63525
ot2:DifficultySum1:ConditionSum1	0.0589004	0.035754	1.64736
ot3:DifficultySum1:ConditionSum1	0.0171475	0.030188	0.56802

maintain the same interpretation as in the original model (the difference between the low and high difficulty versions in the impaired condition relative to the control condition), though their values are reversed (because the condition contrast was reversed) and divided by 4 because the contrast range for each of the two factors was doubled when they went from treatment coding (range: 0 to 1) to sum coding (range: -1 to 1). This difference in the values of the parameter estimates is just a superficial scale difference, which is reflected in the standard errors also being 4 times smaller and the t-values being identical for both kinds of contrast coding.[1] This equivalence is only true for the highest-level interactions because these mean the same thing under either contrast structure. The lower-level effects (difficulty and condition effects) correspond to different comparisons depending on the contrast structure, so these parameter estimates and inferential statistics will be different.

It is perhaps useful to remind ourselves that these changes of reference level and contrast coding only serve to make the parameter estimates more intuitive and do not change the overall model fit. For example, the log likelihoods of our original model and the model with improved factor coding are exactly the same:

```
> logLik(m.ML)

'log Lik.' 1164.8 (df=37)

> logLik(m.MLsum)

'log Lik.' 1164.8 (df=37)
```

Finally, given that the coding of the factor levels and the contrasts determines how the parameter estimates should be interpreted, it is generally a good idea to describe these aspects of the model when reporting the results.

---

[1]A variant of sum coding, called *deviation* coding, avoids this difference by using -0.5 and 0.5 instead of -1 and 1. However, deviation coding is not available as a built-in contrast in R, so it would require hand-coding. This hand-coding is not too difficult, but to keep things simple and to reduce entry points for human errors, we'll use sum coding instead of deviation coding. If you intend to interpret the actual parameter estimate values – not just their direction and statistical significance – then you need to keep the contrast coding scale in mind.

5.3   Multiple comparisons

So far we have considered only categorical predictors that have two levels, with parameters estimated for one level relative to the other. When there are more than two levels, there will be N-1 such relationships, where N is the number of levels and 1 is subtracted for the reference level. To demonstrate how to deal with categorical predictors that have more than two levels let's look at some data from the Moss Aphasia Psycholinguistics Project Database (Mirman et al., 2010). We'll analyze change in proportion of picture naming responses that were semantic errors (such as saying "horse" to a picture of a cow) for a group of aphasic patients. Each patient completed the picture naming test five times over the course of several weeks (see also Schwartz & Brecher, 2000). The numbering of the tests starts from 0 so that the intercept term in the model will capture performance on the first test, which is the baseline assessment. The patients are grouped by aphasia subtype: Anomic (N=6), Conduction (N=9), or Wernicke's (N=8). The data are shown in Figure 5.2.

It looks like the proportion of semantic errors tends to decrease for Anomic aphasic patients, stay about the same for Conduction aphasic patients, and increase for Wernicke's aphasic patients. Let's test this using growth curve analysis. Since we only have five observations per participant, using a higher-order polynomial might overfit the data, so we'll just use a first-order (linear) growth curve analysis and since the first test is a potentially interesting baseline, we'll use a natural polynomial instead of orthogonal polynomial.

```
> m.sem <- lmer(Semantic.error ~ TestTime * Diagnosis +
                               (TestTime | SubjectID),
             data=NamingRecovery, REML=FALSE)
```

To interpret the parameter estimates

```
> coef(summary(m.sem))
```

	Estimate	Std. Error	t value
(Intercept)	0.0457667	0.0077762	5.88546
TestTime	-0.0086850	0.0035242	-2.46440
DiagnosisConduction	-0.0151489	0.0100391	-1.50899
DiagnosisWernicke	-0.0048992	0.0102870	-0.47625
TestTime:DiagnosisConduction	0.0073083	0.0045497	1.60633
TestTime:DiagnosisWernicke	0.0128538	0.0046621	2.75709

we extend the basic logic of treatment coding to three levels: the Anomic group is the reference level, so it is treated as the baseline and parameters are estimated for the Conduction and Wernicke's groups relative to the Anomic group. The (Intercept) is the baseline (*TestTime* = 0) proportion of semantic errors for the Anomic group and TestTime is the slope for the Anomic group. The other parameter estimates comprise a set of pairwise comparisons:

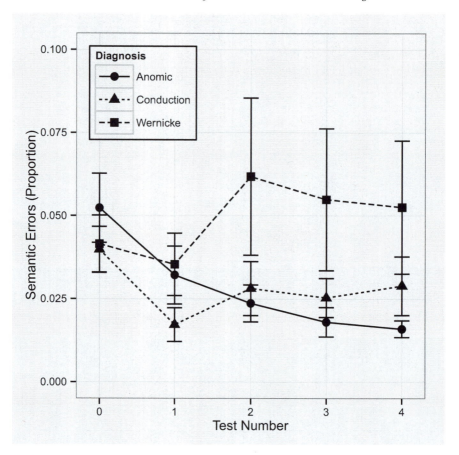

FIGURE 5.2
Proportion of semantic errors in picture naming for three groups of aphasic patients across five test administrations. Error bars indicate ±SE.

`DiagnosisConduction` is the baseline difference between the Anomic and Condition groups, `TestTime:DiagnosisConduction` is the slope difference between the Anomic and Conduction groups, and `DiagnosisWernicke` and `TestTime:DiagnosisWernicke` are the analogous comparisons of the Anomic and Wernicke's groups. From these parameter estimates (and using $t > 2$ as an approximate threshold for $p < 0.05$) we can tell that the Anomic group has a significant decline in semantic errors (*Estimate* = -0.0087, *SE* = 0.0035) and that the difference between the Anomic and Wernicke's groups increases over time (*Estimate* = 0.013, *SE* = 0.0047).

These comparisons give us a lot of information, but do not answer all the questions we might ask. For example, we are missing the pairwise comparison between the Conduction and Wernicke's groups. We also don't know whether the Wernicke's group exhibits a significant overall increase in semantic errors or just an increase relative to the Anomic group (which, remember, is having a decrease in semantic errors). We can answer both of these questions by setting the Wernicke's group to be the reference level for the `Diagnosis` factor

```
> NamingRecovery$DiagnosisW <-
                relevel(NamingRecovery$Diagnosis, "Wernicke")
```

and re-fitting the model.

```
> m.semW <- lmer(Semantic.error ~ TestTime*DiagnosisW +
                            (TestTime | SubjectID),
                data=NamingRecovery, REML=FALSE)
> coef(summary(m.semW))
```

	Estimate	Std. Error	t value
(Intercept)	0.0408675	0.0067344	6.06846
TestTime	0.0041687	0.0030520	1.36589
DiagnosisWAnomic	0.0048992	0.0102870	0.47625
DiagnosisWConduction	-0.0102497	0.0092556	-1.10741
TestTime:DiagnosisWAnomic	-0.0128537	0.0046621	-2.75709
TestTime:DiagnosisWConduction	-0.0055454	0.0041946	-1.32203

This version of the model provides pairwise comparisons between the Wernicke's and Conduction groups (intercept: `DiagnosisWConduction`, slope: `TestTime:DiagnosisWConduction`) and a test of whether the Wernicke's group exhibits a significant increase in semantic errors (it does not: the t-value for the `TestTime` parameter estimate is less than 2).

This re-referencing approach works fine, but it can be very laborious and time-consuming for more complex situations that require multiple re-levelings. Plus, if we look closely at the parameter estimates, we see that all of the information we need was actually present in the original model. Since the Anomic group is a common reference point, we could get the Conduction-Wernicke's comparison (`DiagnosisWConduction`: -0.0102) by computing the difference between the Conduction-Anomic and Wernicke's-Anomic comparisons:
`DiagnosisConduction - DiagnosisWernicke` = -0.0151 - -0.0049 = -0.0102

The `multcomp` package provides a relatively easy way to do these sorts of comparisons using just the original model. First, we need to set up a contrast matrix that defines the comparisons that we want to test. Each column in this matrix corresponds to a parameter estimate from the original model, in the order that they appear in the output. So the first column corresponds to (`Intercept`), the second column to `TestTime`, the fourth column to `DiagnosisWernicke`, etc. Each row in the contrast matrix corresponds to a contrast that we want to test and the elements in the matrix are weights for that contrast. The simplest case is when the contrast we want to test corresponds to an estimated parameter: we put a 1 in that column and a 0 in all of the others. For example, for the slope difference between the Anomic and Wernicke's

groups (which is estimated by `TestTime:DiagnosisWernicke`), we just put a
1 in the sixth column. To test the slope difference between Conduction and
Wernicke's aphasics, we put a -1 in the fifth column and a 1 in the sixth
column. Here is a contrast matrix for testing all pairwise group differences in
this model, with informative labels for each comparison:

```
> contrast.matrix = rbind(
    "Anomic vs. Conduction" = c(0, 0, 1, 0, 0, 0),
    "Anomic vs. Wernicke" = c(0, 0, 0, 1, 0, 0),
    "Conduction vs. Wernicke" = c(0, 0, -1, 1, 0, 0),
    "Slope: Anomic vs. Conduction" = c(0, 0, 0, 0, 1, 0),
    "Slope: Anomic vs. Wernicke" = c(0, 0, 0, 0, 0, 1),
    "Slope: Conduction vs. Wernicke" = c(0, 0, 0, 0, -1, 1))
```

Now we can use this contrast matrix to test the pairwise comparisons. First,
we load the `multcomp` package:

```
> library(multcomp)
```

Then we use the `glht` function to compute the pairwise comparisons defined
by the contrast matrix within the context of the model (`m.sem`):

```
> comps <- glht(m.sem, contrast.matrix)
```

We can use the `summary` function to get a list of the parameter estimates,
standard errors, z-values, and *p*-values for each of those comparisons:

```
> summary(comps)
          Simultaneous Tests for General Linear Hypotheses

Fit: lmer(formula = Semantic.error ~ TestTime * Diagnosis +
    (TestTime | SubjectID), data = NamingRecovery, REML = FALSE)

Linear Hypotheses:
                                  Estimate Std. Error z value
Anomic vs. Conduction == 0        -0.01515    0.01004   -1.51
Anomic vs. Wernicke == 0          -0.00490    0.01029   -0.48
Conduction vs. Wernicke == 0       0.01025    0.00926    1.11
Slope: Anomic vs. Conduction == 0  0.00731    0.00455    1.61
Slope: Anomic vs. Wernicke == 0    0.01285    0.00466    2.76
Slope: Conduction vs. Wernicke == 0 0.00555   0.00419    1.32
                                  Pr(>|z|)
Anomic vs. Conduction == 0          0.490
Anomic vs. Wernicke == 0            0.986
Conduction vs. Wernicke == 0        0.759
Slope: Anomic vs. Conduction == 0   0.426
Slope: Anomic vs. Wernicke == 0     0.032 *
Slope: Conduction vs. Wernicke == 0 0.618
```

```
---
Signif. codes:  0 '***' 0.001 '**' 0.01 '*' 0.05 '.' 0.1 ' ' 1
(Adjusted p values reported -- single-step method)
```

By default, the *p*-values are adjusted, that is, corrected for multiple comparisons. A number of different corrections are available, including no correction (just the *p*-values estimated from the normal distribution as we've done before):

```
> summary(comps, test = adjusted("none"))
         Simultaneous Tests for General Linear Hypotheses

Fit: lmer(formula = Semantic.error ~ TestTime * Diagnosis +
    (TestTime | SubjectID), data = NamingRecovery, REML = FALSE)

Linear Hypotheses:
                                   Estimate Std. Error z value
Anomic vs. Conduction == 0         -0.01515    0.01004   -1.51
Anomic vs. Wernicke == 0           -0.00490    0.01029   -0.48
Conduction vs. Wernicke == 0        0.01025    0.00926    1.11
Slope: Anomic vs. Conduction == 0   0.00731    0.00455    1.61
Slope: Anomic vs. Wernicke == 0     0.01285    0.00466    2.76
Slope: Conduction vs. Wernicke == 0 0.00555    0.00419    1.32
                                   Pr(>|z|)
Anomic vs. Conduction == 0           0.1313
Anomic vs. Wernicke == 0             0.6339
Conduction vs. Wernicke == 0         0.2681
Slope: Anomic vs. Conduction == 0    0.1082
Slope: Anomic vs. Wernicke == 0      0.0058 **
Slope: Conduction vs. Wernicke == 0  0.1862
---
Signif. codes:  0 '***' 0.001 '**' 0.01 '*' 0.05 '.' 0.1 ' ' 1
(Adjusted p values reported -- none method)
```

5.4   Chapter recap

This chapter provided a basic guide to handling categorical predictors. Section 5.2 explained how a categorical predictor is converted to numeric values for the purposes of regression. The default approach is to consider the reference level of the categorical variable to be the baseline and estimate parameters for each of the other levels relative to this baseline. This is called *treatment* or *dummy* coding. Treatment coding is simple and works well in many cases, but it can produce parameter estimates that are difficult or counterintuitive to interpret. This is particularly true for factorial designs that have more than one categorical predictor, in which case the parameter estimates correspond to simple effects rather than main effects. *Sum* or *deviation* coding, in which the levels are coded as opposite deviations from an intermediate baseline, provides a useful alternative that produces more easily interpretable parameter estimates.

Section 5.3 described two ways to conduct multiple pairwise comparisons among levels of a categorical predictor variable that has more than two levels. The first was to fit multiple versions of the same model setting different levels of the factor as the baseline. This approach is easy to implement, but it is repetitive, can be time-consuming, and has more opportunity to introduce typo or copy-and-paste errors. An alternative is to fit just one model and build a contrast matrix that defines the comparisons of interest, then use the `glht` function from the `multcomp` package to evaluate those comparisons. This function also provides access to built-in corrections for multiple comparisons, although those corrections may or may not be desired or appropriate.

Changes of contrast coding and reference levels affect the parameter estimates, but not the overall model fit. In other words, they may change what the model tells you about your data, but not how well the model fits the data. Also, keep in mind that these issues apply to regression in general, not just growth curve analysis or multilevel regression, so the topics covered in this chapter may prove useful for other regression analyses.

5.5   Exercises

1. The `FunctTheme` data frame contains data on the time course of activation of thematic and function relationships (discussed in Chapter 4). The study was a factorial design with two categorical predictor variables: `Condition` (`Function` vs. `Thematic`) and `Object` (related `Competitor` vs. `Unrelated` distactor; omit the `Target` object for this exercise).

   (a) Analyze the competition data using GCA with fourth-order orthogonal polynomials and the default treatment coding of the factors. Interpret the parameter estimates – explicitly identify which aspect of the data or comparison is captured by each parameter estimate.

   (b) Re-code the factors to make the parameter estimates more intuitive to interpret: set the `Unrelated` object to be the reference level for the `Object` factor and change the contrast coding for the `Condition` factor from *treatment* to *sum*. Fit a new GCA model and explain any changes in the parameter estimates.

2. Use the full model of the WISQARS suicide data to test all pairwise comparisons between regions.

   (a) Which regions differ from which other regions in terms of baseline suicide rate?

   (b) Which regions differ from which other regions in terms of rate of change of suicide rate?

   (c) Re-analyze the data with orthogonal time to estimate differences in overall suicide rate instead of "baseline" differences in 1999. Are any of the results different?

6

Binary outcomes: Logistic GCA

CONTENTS

6.1   Chapter overview

So far, all of the examples have treated the outcome variables as continuous – as if the outcome variable could hypothetically take any value. This is approximately true for variables such as reaction time or income, but it is not true for variables such as accuracy, which can only take a specific set of discrete values, such as "correct" or "incorrect." This chapter will begin by describing the problems that can arise from treating such *binary outcome variables* (also called *dichotomous variables*) as if they are continuous.

It will then describe two *logistic* extensions of the basic GCA approach that appropriately treat binary variables. The first applies multilevel logistic regression directly to the binary outcome data. The second uses the *empirical logit* transformation to rescale the binary outcome data to a continuous variable and then applies the same linear approach described in the previous chapters. The chapter will conclude with a demonstration of how to plot model fits from logistic GCA.

6.2   Why binary outcomes need logistic analyses

In the psychological and neural sciences we typically treat outcome (or response) variables as continuous – as if they could have any value. However, this is often not true: in any binary choice task (such as answering yes-no questions, word-to-picture matching, lexical decision, fixating one object vs. another, and many others), the outcome on a given trial can only take one of two values. The same is true when more open-ended tasks, such as picture naming or problem solving, are made binary by scoring responses as correct vs. incorrect. Binary variables follow a *binomial* distribution and have two properties that make standard linear statistics inappropriate.

The first is that the range of possible sample means (e.g., proportion of correct responses) is bounded between 0 and 1. Linear statistics do not include these bounds in the analysis, which can lead to spurious and uninterpretable results (such as predicting accuracies greater than 1 or lower than 0). Also, these floor and ceiling bounds can make data look asymptotic (i.e., the data tend to plateau near 0 and 1), which can seem like justification for fitting sigmoid or other asymptotic functions instead of polynomial functions. However, binary data can be asymptotic for two very different reasons. If the data are asymptotic because the underlying psychological processes are asymptotic, then a non-linear function might be justified (though recall the discussion in Chapter 3 about the dangers involved in choosing such a function). On the other hand, the data might be asymptotic because the nature of the task and measurement creates floor and ceiling effects (as would happen in a binary choice task), not because the underlying process is asymptotic. In this case, logistic regression is the right solution to the problem of asymptotic data.

The second issue is that the variance is not constant (*homogeneous*) over the range of possible proportions: it is larger near the middle (0.5) than near the ends (1.0 or 0.0). To get an intuitive sense of why this is true, imagine flipping a fair coin 10 times. The central tendency is to get 5 "heads" and 5 "tails" outcomes, but of course sometimes you will get 4 "heads" or 6 "heads" or maybe even only 2 or 3 "heads." In other words, there will be a lot of variability. Now imagine flipping a very biased coin that only comes up "heads" 5% of the time. In a sample of 10 flips, you will probably only have 0 or 1 "heads," maybe 2 "heads" – there will be much less variability in the outcome. This property is illustrated in Figure 6.1, which shows the binomial distribution's variance at different levels of sample proportions (i.e., probability of "heads") for a few different sample sizes (i.e., number of coin flips). Although the variance distortion gets smaller for larger sample sizes, it is an intrinsic property of the binomial distribution.

The larger variance in the middle of the range means that, for any observed sample, differences in the middle of the range are less reliable than differences near the ends. This non-homogeneity of variance in binomial data can cause linear analysis methods to produce incorrect results, both false positives and false negatives. To demonstrate this, imagine a treatment or training study that tested three groups of 15 participants. The groups differ in the initial severity of their impairment. The participants complete a 30-trial pretest before the intervention and then complete it again after the intervention (posttest). The fictitious data are shown in Figure 6.2. The mild severity

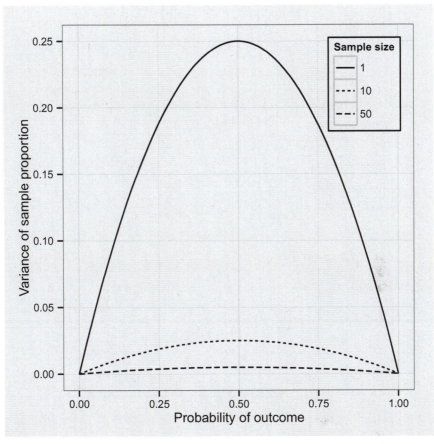

FIGURE 6.1
Variance of sample proportion as a function of true probability of outcome and sample size.

group improved from about 90% correct to about 96% correct, the moderate severity group improved from about 60% correct to about 85% correct, and the most severely impaired group improved from about 5% correct to about 31% correct. On this linear percent-correct scale, the mild group shows the smallest improvement (about 6%) and the moderate and severe groups show about the same amount of improvement (about 25%). The linear regression results for group differences in amount of change from pretest to posttest (Table 6.1) reflect this pattern: the mild group showed a smaller change than the moderate and severe groups, which were nearly identical to one another.

However, the moderate group's performance was closer to the middle of the range, so it should be more variable, so maybe that difference is less meaningful. Indeed, a logistic regression (Table 6.2) reveals a very different pattern: the change from pretest to posttest was not significantly different

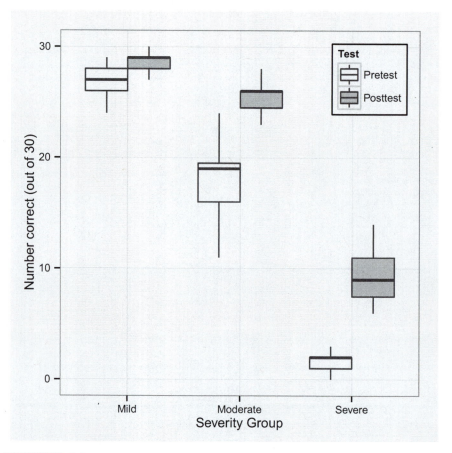

FIGURE 6.2
Pretest and posttest performance for each group.

TABLE 6.1
Change from Pretest to Posttest: Linear Regression Results

	Estimate	Std. Error	z value	$\Pr(>|z|)$
Mild vs. Moderate	5.800	0.958	6.052	0.000
Mild vs. Severe	5.867	0.958	6.121	0.000
Moderate vs. Severe	0.067	0.958	0.070	0.945

between the mild and moderate groups, but it was significantly larger for the severe group. In other words, getting 7 or 8 more correct responses out of 30 trials (about 25%) means a lot more when you start out getting only 1 or 2 correct than when you start out getting 18 correct. Conversely, getting 2 more correct responses when you were already getting 27 right out of 30 means as much as getting 7 or 8 more when you started in the middle of the range.

TABLE 6.2

Change from Pretest to Posttest: Logistic Regression Results

	Estimate	Std. Error	z value	Pr($>|z|$)
Mild vs. Moderate	0.347	0.333	1.043	0.297
Mild vs. Severe	1.113	0.375	2.973	0.003
Moderate vs. Severe	0.766	0.291	2.633	0.008

In sum, binary outcome data follow a binomial distribution, which is bounded at proportions of 0 and 1, and has systematically non-homogeneous variance. These properties can produce spurious results, both false positives and false negatives. Logistic regression correctly models these properties and provides more reliable results for analyses of binary outcomes (for a more mathematically detailed discussion with additional examples, see Jaeger, 2008). The next two sections will discuss and demonstrate how to perform logistic growth curve analyses.

6.3   Logistic GCA

Logistic regression models the binomial process that produces binary data, so the outcome variable in the data set needs to be those binary data. In other words, it is not sufficient to know that a particular participant was 90% correct. The model needs to know whether that 90% was 9 out of 10 trials or 90 out of 100 trials. This information can be provided as a binary vector of 0's and 1's where each value corresponds to a single trial, or, more compactly, as counts of the number of "successes" and the number of "failures" or number of trials.

In Chapter 4 we used linear GCA to analyze fixation data on the time course of recognition for high and low frequency words. Fixation is a binary variable — at each point in time, participants either fixate the target or they don't — so let's revisit those data with logistic GCA.

```
> summary(TargetFix)
    Subject         Time            timeBin     Condition
 708    : 30   Min.   : 300   Min.   : 1    High:150
 712    : 30   1st Qu.: 450   1st Qu.: 4    Low :150
 715    : 30   Median : 650   Median : 8
 720    : 30   Mean   : 650   Mean   : 8
 722    : 30   3rd Qu.: 850   3rd Qu.:12
 725    : 30   Max.   :1000   Max.   :15
 (Other):120
    meanFix             sumFix              N
```

```
Min.    :0.0286    Min.    : 1.0    Min.    :33.0
1st Qu.:0.2778    1st Qu.:10.0    1st Qu.:35.8
Median :0.4558    Median :16.0    Median :36.0
Mean    :0.4483    Mean    :15.9    Mean    :35.5
3rd Qu.:0.6111    3rd Qu.:21.2    3rd Qu.:36.0
Max.    :0.8286    Max.    :29.0    Max.    :36.0
```

As discussed in Chapter 4, participants were asked to pick which one of four pictures matched a spoken word and the gradual rise in fixations on the target picture reveals the gradual comprehension of the spoken word. In Chapter 4, the outcome variable was `meanFix`: the proportion of trials on which the target picture was fixated by each participant, in each condition, in each time bin. Logistic GCA will use the numerator and denominator from computing that proportion: `N` is the number of trials for each participant in each condition (that is, the number of opportunities to fixate the target picture) and `sumFix` is the number of trials on which the target was fixated by each participant in each condition in each time bin.

As in Chapter 4, we prepare the data for analysis by creating a third-order orthogonal polynomial

```
> t <- poly(unique(TargetFix$timeBin), 3)
```

and appending the orthogonal polynomial values to the `TargetFix` data frame, aligned with their corresponding time bins.

```
> TargetFix[,paste("ot", 1:3, sep="")] <-
                              t[TargetFix$timeBin, 1:3]
```

The model syntax will be very similar to the linear GCA in Chapter 4, with two critical differences. First, the outcome variable will be a pair of columns: the counts of successes (`sumFix`) and the counts of failures (`N − sumFix`). This pair will be created by using the function `cbind` (column **bind**). Second, we need to use the *generalized* form of multilevel regression, so we use the `glmer` function and specify that the error distribution should be of the binomial family. Here is the syntax for the full model:

```
> m.log <- glmer(cbind(sumFix, N-sumFix) ~
                        (ot1+ot2+ot3)*Condition +
                        (ot1+ot2+ot3 | Subject) +
                        (ot1+ot2 | Subject:Condition),
                data=TargetFix, family=binomial)
```

Compared to linear models, logistic models are more susceptible to convergence failures with complex random effect structures. In fact, the full logistic GCA model for these data did not converge and it was necessary to simplify the `Subject:Condition` random effect by removing the cubic term. Logistic models also take substantially longer to fit than linear models do (for this example, it was about 10 times longer).

When examining the model results, the `summary` output for a logistic GCA model will automatically use the normal distribution to estimate *p*-values for the fixed effects (Table 6.3). The results from a linear GCA model that used `meanFix` as the outcome variable and the full random effect structure are shown in Table 6.4. The parameter estimates are quite different between the two models because the outcomes are on different scales (linear vs. logistic), though in this case the interpretation of the Condition effects would be largely the same: there were significant effects of Condition on the intercept and on the quadratic term, indicating faster word recognition for high frequency words than low frequency words.

TABLE 6.3
Target Fixation: Logistic GCA Results

	Estimate	Std. Error	z value	$Pr(>|z|)$
(Intercept)	-0.117	0.065	-1.785	0.074
ot1	2.819	0.298	9.457	0.000
ot2	-0.559	0.169	-3.306	0.001
ot3	-0.321	0.127	-2.520	0.012
ConditionLow	-0.262	0.091	-2.877	0.004
ot1:ConditionLow	0.064	0.331	0.194	0.846
ot2:ConditionLow	0.695	0.239	2.903	0.004
ot3:ConditionLow	-0.071	0.166	-0.425	0.670

TABLE 6.4
Target Fixation: Linear GCA Results

	Estimate	Std. Error	*t*	*p*
(Intercept)	0.477	0.014	34.458	0.000
ot1	0.639	0.060	10.654	0.000
ot2	-0.110	0.038	-2.848	0.004
ot3	-0.093	0.023	-4.002	0.000
ConditionLow	-0.058	0.019	-3.093	0.002
ot1:ConditionLow	0.000	0.066	0.005	0.996
ot2:ConditionLow	0.164	0.054	3.033	0.002
ot3:ConditionLow	-0.002	0.027	-0.077	0.938

6.4 Quasi-logistic GCA: Empirical logit

As we've seen, small differences near the endpoints (0 and 1) can have very big impacts on a logistic scale. Because studies have a finite (and often relatively small) number of trials, there is some granularity to the observed proportions.

For example, in a study with 20 trials, it is only possible to observe values of 100% correct, 95% correct, 90% correct, etc., but the logistic scale difference between those values is very large. Logistic regression relies on the *logit* or *log-odds* transformation, which is given in Equation 6.1, where Y is the number of "successes" and N is the number of trials (i.e., $p = Y/N$).

$$logit(Y, N) = \log\left(\frac{Y}{N - Y}\right) \tag{6.1}$$

This function is undefined when $Y = N$ (i.e., $p = 1$) because it would mean dividing by 0, and when $Y = 0$ (i.e., $p = 0$) because $\log(0)$ is undefined. So in addition to problems due to low resolution near the boundaries, perfect scores (i.e., $p = 0$ or $p = 1$) can undermine the accuracy of logistic regression.

One solution to these problems is to use *quasi*-logistic regression. In particular, a useful alternative is the *empirical logit transformation*, which employs a logistic scale but incorporates an adjustment for the granularity of the data (for more discussion and examples, see Barr, 2008). The empirical logit transformation (Equation 6.2) adds a 0.5 adjustment factor, which avoids both of the undefined boundary conditions and scales with the number of observations (the adjustment becomes functionally smaller as the number of observations increases).

$$elogit(Y, N) = \log\left(\frac{Y + 0.5}{N - Y + 0.5}\right) \tag{6.2}$$

Figure 6.3 shows the logit transformation (solid line) along with the empirical logit approximation at a relatively large number of trials ($N = 100$, dashed line) and a relatively small number of trials ($N = 10$, dotted line). When the number of trials is relatively large, the empirical logit very closely approximates the logit function, but when the number of trials is relatively small, the very steep change near the endpoints is mitigated. This is exactly the behavior we want if we're concerned that the small number of trials might make our estimates of p unreliable near the endpoints.

To use the empirical logit in GCA, we first need to compute the empirical logit transformation from the observed "successes" and number of trials. Returning to the fixation data, we'll compute the empirical logit:

```
> TargetFix$elog <- with(TargetFix,
                         log((sumFix+0.5) / (N-sumFix+0.5)))
```

The `with` function takes a data frame and an expression, and evaluates the expression in the *environment* of the data frame. This is a convenient shortcut and makes the code a little easier to read: instead of typing `TargetFix$N-TargetFix$sumFix+0.5` you can just type `N-sumFix+0.5`.

We can further improve the model by assigning *weights* to observations based on their reliability. That is, we can explicitly tell the model that empirical logit values that are based on fewer trials and that are closer to the

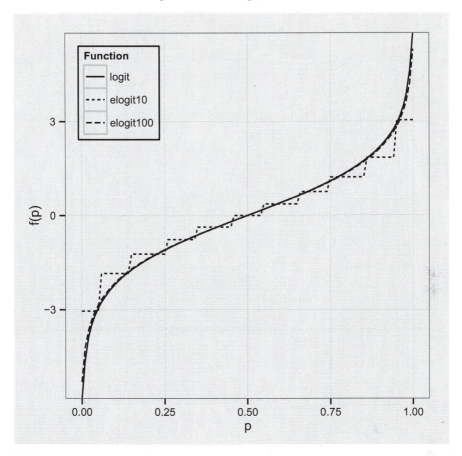

FIGURE 6.3
Logit transformation (solid line) and the empirical logit approximation at
N=100 (dashed line) and N=10 (dotted line).

endpoints are less reliable and should have less weight in the regression. To
do this, we compute weights w based on the variance of the empirical logit
function as given in Equation 6.3:

$$w(Y, N) = \frac{1}{Y + 0.5} + \frac{1}{N - Y + 0.5} \tag{6.3}$$

In our fixation data set, implementing that equation will look like this:

```
> TargetFix$wts <- with(TargetFix,
                  1/(sumFix+0.5) + 1/(N-sumFix+0.5))
```

Now we can fit a standard linear GCA model using `elog` as the outcome
variable and the inverse of `wts` as the weights (so the less reliable near-endpoint
values will have less weight):

```
> m.elog <- lmer(elog ~ (ot1+ot2+ot3)*Condition +
                        (ot1+ot2+ot3 | Subject) +
                        (ot1+ot2+ot3 | Subject:Condition),
                  control=lmerControl(optimizer="bobyqa"),
                  data=TargetFix, weights=1/wts, REML=F)
```

As with the linear GCA on fixation proportions (meanFix), the empirical logit GCA can handle the full random effects structure, the computation is faster, and the results do not (by default) include *p*-values. The empirical logit model results are shown in Table 6.5. As in the other two models, there are strong effects of Condition on the intercept and quadratic terms. Notice also that the parameter estimates are fairly similar to the logistic GCA results in Table 6.3 because the data are now on the same (logistic) scale, though the parameter estimates are not exactly the same because the empirical logit is only an approximation of the true logit values computed by the logistic regression.

TABLE 6.5
Target Fixation: Empirical Logit GCA Results

	Estimate	Std. Error	t	p
(Intercept)	-0.113	0.022	-5.062	0.000
ot1	2.725	0.102	26.798	0.000
ot2	-0.545	0.058	-9.334	0.000
ot3	-0.303	0.037	-8.098	0.000
ConditionLow	-0.248	0.031	-8.010	0.000
ot1:ConditionLow	0.032	0.113	0.288	0.773
ot2:ConditionLow	0.689	0.082	8.430	0.000
ot3:ConditionLow	-0.084	0.045	-1.855	0.064

6.5   Plotting model fits

The fitted function returns estimated probabilities even for logistic models, so plotting model fits for logistic GCA models is essentially the same as plotting model fits for linear GCA models. Figure 6.4 shows the observed data and model fits for the three different versions of GCA discussed in this chapter, generated using code of the form

```
> ggplot(TargetFix, aes(Time, Observed, shape=Condition)) +
    stat_summary(fun.data=mean_se, geom="pointrange") +
    stat_summary(aes(y=Fit, linetype=Condition),
                 fun.y=mean, geom="line") +
    ylab("Outcome") + theme_bw(base_size=10)
```

where `Observed` is the name of the variable with the observed data and `Fit` is the name of the variable with the model fit.

6.6 Chapter recap

This chapter covered how to deal with binary outcome variables. Linear statistics, including ANOVA, linear regression, and GCA methods from previous chapters, are not appropriate for binary outcome variables because the range of possible outcome values is bounded at 0 and 1 and because the variance is not constant over that range. Logistic regression provides a way to appropriately model binary data, including capturing asymptotic patterns that arise due to floor and ceiling effects. The syntax for logistic GCA is very similar to the syntax for linear GCA. This chapter covered two possible implementations: logistic GCA and quasi-logistic GCA using empirical logits.

Logistic GCA can be implemented by specifying the outcome variable as either a binary column of 0's and 1's or a pair of columns containing the count of "successes" (number of 1's) and the count of "failures" (number of 0's), and the distribution family as `binomial`. This approach is the true logistic regression approach adapted to GCA. The downside of this approach is that it takes substantially longer to fit such models and they have more difficulty with convergence, which may require simplifying the random effects.

Also, when the number of trials is not very large, the granularity of the data will cause values close to the bounds to be problematic for true logistic regression. In this case, it may be more effective to take a quasi-logistic approach implemented using the empirical logit transformation. The empirical logit approximates the true logit, but incorporates an adjustment that avoids the boundary condition problems. This approach also has the benefits of faster computation and being able to handle more complex random effect structures.

6.7 Exercises

The word learning accuracy data in `WordLearnEx` are proportions from a binary response variable (correct/incorrect). Re-analyze these data using logistic and quasi-logistic GCA and compare the results to linear GCA from Chapter 3.

1. Convert the accuracy proportions to number of correct and incorrect responses (there were 6 trials per block). Compute the empirical logits and corresponding weights.

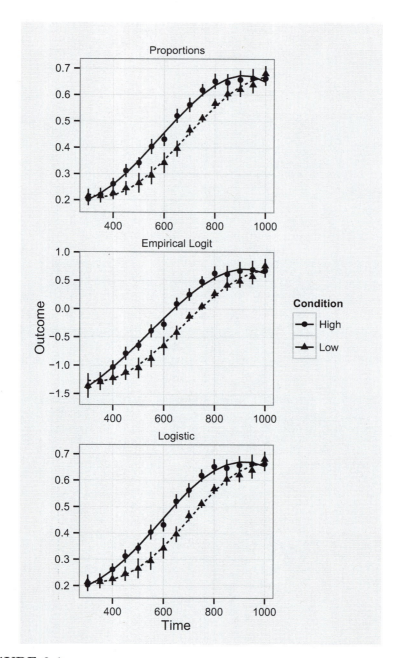

FIGURE 6.4
Observed data (symbols, vertical lines indicate ±SE) and GCA model fits
(lines) for three different versions of GCA.

2. Analyze the data using logistic GCA. Are the results different from linear GCA? If so, how and why?

3. Analyze the data using quasi-logistic GCA with empirical logits. Are the results different from linear GCA? If so, how and why? Are they different from logistic GCA? If so, how and why?

4. Plot the model fits for each of the analyses.

7

Individual differences

CONTENTS

7.1 Chapter overview

Every researcher that has run a study with human (or non-human animal) participants has seen individual differences. Some participants show a large effect, some show no effect; some show an early effect, some show a later effect, etc. There are two ways of thinking about this variability. The first is to treat all of this variability as random noise and simply ask whether there is some signal (a statistically reliable effect) present within that noise. There is no doubt that some amount of random noise is present in all behavioral data: even in the simplest reaction time experiments, screen refresh rates and other computer timing issues will introduce a small amount of measurement noise into the data. A *t*-test or ANOVA treats variability as random noise that serves as the background for estimating the reliability of the signal. However, it is also possible that some of the variability we observe is not completely random, that it is systematically related to properties of the individual participants. Such systematic effects should have implications for the theory being tested. In other words, whatever phenomenon or process is being studied, systematic individual differences provide additional insights into that phenomenon or process. Growth curve analysis (and multilevel regression more generally) provides a statistical tool to quantify systematic individual differences, which can then be used to extract new insights. How to do that is the subject of this chapter.

The first method will be familiar from other regression-based approaches: if the individual difference variable is known and measured in advance (IQ, age, impairment severity, etc.), then it can be added to the growth curve model as a fixed effect. Like other fixed effects, it can be added as an effect on specific time terms (intercept, linear, quadratic, etc.) to evaluate individual differences in particular aspects of the growth trajectory. Section 7.2 will discuss and demonstrate how to apply this method of individual difference analysis in the context of a linear growth curve analysis.

The second method is to use the random effects as estimates of how each participant differs from the overall group pattern. This approach is useful when the individual differences of interest are internal to the study itself (i.e., there is no external measure that could be entered as a fixed effect) or when the individual differences need to be quantified for an analysis that is impossible or impractical to integrate with GCA (for example, neuroimaging analyses). Section 7.3 will briefly review random effects (Chapters 2 and 4 covered random effects in more detail) and demonstrate how to extract and manipulate random effects from a growth curve model and use them to quantify individual differences.

7.2   Individual differences as fixed effects

The simplest approach to analyzing individual differences is just to treat them the same way we have treated study manipulations throughout this book: by adding them as fixed effects to the model. This approach works well when the individual differences are based on some assessment that is outside of the data being analyzed. For example, the time course of learning to perform some cognitive task (for example, learning a set of new words) might be influenced by participant age or working memory span or severity of language impairment. In this type of case, in addition to the experimental task, we could administer a test of working memory span (or language impairment severity, etc.) and include that score as a fixed effect in the growth curve model of word learning.

Two things are important to keep in mind when taking this approach. First, participants can be randomly assigned to experimental conditions, but their individual differences are not randomly assigned, so the results will always be subject to "third variable" explanations. That is, rather than the measured individual differences directly causing the differences on the experimental task, there may be some third variable that is responsible for both the individual differences and the variation in experimental task performance. Second, the individual difference predictors can be individual-level continuous variables (for example, each participant's unique *age*) or they can be group-level categorical variables (for example, *children* vs. *adults*), but the latter will be subject to power issues if the groups are small. One of the advantages of using regression-based methods over ANOVAs is that ANOVAs only handle categorical predictors, but regression modeling allows continuous predictors, which often have more power to detect differences.

In general, many considerations that are familiar from multiple regression methods will be relevant to testing individual differences using fixed effects, including continuous vs. categorical predictors and concerns about collinearity between predictors. The key difference is that in multilevel (growth curve) models, individual-level predictors are tested for their effect on the set of observations corresponding to that individual. Let's work through an example using a subset of data from a longitudinal study of reading development (from the ELDEL project, see Caravolas et al., 2012, 2013, and http://www.eldel.eu/. Thanks to Markéta Caravolas, Charles Hulme, and the ELDEL team for sharing these data).

We will analyze data from 181 children learning to read English. The children were tested six times, approximately at the middle and end of each school year for three years, starting from reception year (kindergarten, when they were approximately five years old) until the end of second grade. The *Time* variable will be represented as months from Time 1 (baseline) testing so that the intercept term will correspond to baseline performance. At each of these six test times, each child's reading ability was assessed by showing them a picture and four printed words and asking them to pick the word that corresponds to the picture ("picture-word matching," `pwmcor`). Figure 7.1 shows the overall improvement in picture-word matching performance (i.e., learning to read), which followed an approximately linear trajectory.

In addition to the measure of reading ability, at Time 1, children were tested on four other measures that may be related to learning to read. These are the measures of individual differences that will be tested to examine the cognitive factors that predict development of reading:

Verbal memory span (`wdspan1`) Children were asked to repeat, in order, lists of familiar words.

Letter knowledge (`lk1`) Children were asked to pronounce the sounds and names of each letter of the English alphabet.

Rapid automatized naming (`ran1`) Children were asked to name, as quickly as they could, five familiar items (objects or colors) and to repeat this list eight times.

Phoneme awareness (`pa1`) Children were asked to pronounce the first or last sound in fake words and to blend two segments into a word.

Here is a summary of the data set:

```
> summary(ELDEL)
      id            wdspan1            lk1              ran1
 ABBTUS :   6   Min.   :1.00    Min.   :-1.450   Min.   :-1.687
```

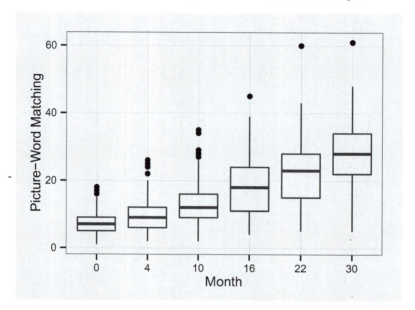

FIGURE 7.1
Overall development of reading ability.

```
ABIARM :    6    1st Qu.:2.00    1st Qu.:-0.114    1st Qu.:-0.623
ABIHAR :    6    Median :3.00    Median : 0.343    Median :-0.115
ABIJON :    6    Mean   :2.61    Mean   : 0.409    Mean   : 0.196
AIDGRI :    6    3rd Qu.:3.00    3rd Qu.: 0.941    3rd Qu.: 0.802
AISELE :    6    Max.   :4.00    Max.   : 2.102    Max.   : 3.684
(Other):1050
      pa1                  pwmcor            Month
 Min.   :-1.39468    Min.   : 1.0     Min.   : 0.0
 1st Qu.:-0.00416    1st Qu.: 8.0     1st Qu.: 4.0
 Median : 0.62316    Median :13.0     Median :13.0
 Mean   : 0.52247    Mean   :15.9     Mean   :13.7
 3rd Qu.: 1.04767    3rd Qu.:23.0     3rd Qu.:22.0
 Max.   : 1.88485    Max.   :61.0     Max.   :30.0
                     NA's   :82
```

Not surprisingly, the individual difference measures were correlated with one
another, particularly letter knowledge and phoneme awareness:

```
> cor(subset(ELDEL, Month == 0,
            select=c("wdspan1", "lk1", "ran1", "pa1")))
        wdspan1      lk1     ran1      pa1
wdspan1 1.00000  0.32789 -0.20032  0.37344
lk1     0.32789  1.00000 -0.35685  0.66537
```

```
ran1     -0.20032 -0.35685  1.00000 -0.40747
pa1       0.37344  0.66537 -0.40747  1.00000
```

As in multiple regression more generally, this means that the estimated effect of any one predictor will depend on which other predictors are included in the model. For example, the effect of letter knowledge will be different depending on whether phoneme awareness is or is not included in the model. Let's begin by building a base model that just describes the overall development of reading skill:

```
> eldel.base <- lmer(pwmcor ~ Month + (Month | id),
                data=ELDEL, REML=FALSE)
```

We can test whether letter knowledge at Time 1 was related to initial reading ability by adding an effect of lk1 on the intercept (recall that Time 1 testing was at Month 0, so this corresponds to the intercept):

```
> eldel.lk <- lmer(pwmcor ~ Month + lk1 + (Month | id),
                data=ELDEL, REML=FALSE)
```

and using the anova function to test whether there was a significant improvement in model fit:

```
> anova(eldel.base, eldel.lk)
Data: ELDEL
Models:
eldel.base: pwmcor ~ Month + (Month | id)
eldel.lk: pwmcor ~ Month + lk1 + (Month | id)
           Df  AIC  BIC logLik deviance Chisq Chi Df Pr(>Chisq)
eldel.base  6 5863 5893  -2926     5851
eldel.lk    7 5818 5852  -2902     5804  47.5      1    5.6e-12

eldel.base
eldel.lk    ***
---
Signif. codes:  0 '***' 0.001 '**' 0.01 '*' 0.05 '.' 0.1 ' ' 1
```

There was a very significant improvement in model fit and examining the parameter estimates

```
> coef(summary(eldel.lk))
            Estimate Std. Error t value
(Intercept) 5.85898   0.258831 22.6363
Month       0.68128   0.024079 28.2932
lk1         2.36959   0.285090  8.3117
```

shows that the estimate for the lk1 term was positive – children with better letter knowledge at baseline were also better at reading at baseline. The same approach can be used to assess the effect of phoneme awareness:

```
> eldel.pa <- lmer(pwmcor ~ Month + pa1 + (Month | id),
                    data=ELDEL, REML=FALSE)
> anova(eldel.base, eldel.pa)
Data: ELDEL
Models:
eldel.base: pwmcor ~ Month + (Month | id)
eldel.pa: pwmcor ~ Month + pa1 + (Month | id)
           Df  AIC  BIC logLik deviance Chisq Chi Df Pr(>Chisq)
eldel.base  6 5863 5893  -2926     5851
eldel.pa    7 5821 5856  -2904     5807  43.9      1    3.4e-11

eldel.base
eldel.pa   ***
---
Signif. codes:  0 '***' 0.001 '**' 0.01 '*' 0.05 '.' 0.1 ' ' 1
```

Like letter knowledge, phoneme awareness was strongly related to initial reading ability and examining the parameter estimates reveals that the effect was
similarly positive:

```
> coef(summary(eldel.pa))
            Estimate Std. Error t value
(Intercept) 5.58944    0.282083 19.8149
Month       0.68084    0.024113 28.2349
pa1         2.36917    0.307314  7.7093
```

Indeed, the two parameter estimates were nearly identical (lk1: 2.3696, pa1:
2.3692), which is a good reminder that phoneme awareness and letter knowledge were strongly positively correlated ($r = 0.665$). That is, both phoneme
awareness and letter knowledge were positively related to reading ability, but
it is hard to know how much of those effects is due to their *shared* contributions and how much is due to their *unique* contributions. To assess the unique
contributions, we need to compare the effect of adding each term to a model
that already has the other term. Since we already have models with the two
effects individually, we just need to build a combined model that has both
effects:

```
> eldel.lk.pa <- lmer(pwmcor ~ Month + lk1 + pa1 +
                      (Month | id), data=ELDEL, REML=FALSE)
```

Now we can evaluate the unique contribution of letter knowledge by comparing
this combined model to one that only has phoneme awareness:

```
> anova(eldel.pa, eldel.lk.pa)
Data: ELDEL
Models:
eldel.pa: pwmcor ~ Month + pa1 + (Month | id)
```

```
eldel.lk.pa: pwmcor ~ Month + lk1 + pa1 + (Month | id)
            Df  AIC  BIC loglik deviance Chisq Chi Df Pr(>Chisq)
eldel.pa     7 5821 5856  -2904     5807
eldel.lk.pa  8 5808 5847  -2896     5792  15.7      1     7.5e-05

eldel.pa
eldel.lk.pa ***
---
Signif. codes:  0 '***' 0.001 '**' 0.01 '*' 0.05 '.' 0.1 ' ' 1
```

The effect of letter knowledge still highly significantly improves model fit, though notice that the improvement is much smaller than when phoneme awareness was not already included (-2LL: 16 vs. 47). We can use the analogous comparison to test the unique effect of phoneme awareness:

```
> anova(eldel.lk, eldel.lk.pa)
Data: ELDEL
Models:
eldel.lk: pwmcor ~ Month + lk1 + (Month | id)
eldel.lk.pa: pwmcor ~ Month + lk1 + pa1 + (Month | id)
            Df  AIC  BIC loglik deviance Chisq Chi Df Pr(>Chisq)
eldel.lk     7 5818 5852  -2902     5804
eldel.lk.pa  8 5808 5847  -2896     5792  12.2      1     0.00049

eldel.lk
eldel.lk.pa ***
---
Signif. codes:  0 '***' 0.001 '**' 0.01 '*' 0.05 '.' 0.1 ' ' 1
```

The effect of phoneme awareness on baseline reading ability significantly improved model fit even when letter knowledge was already in the model, though again the improvement was substantially smaller (-2LL: 12 vs. 44). These comparisons indicate that phoneme awareness and letter knowledge made both shared and unique contributions to baseline reading ability.

If we are only interested in some of the predictor variables and simply want to control for the remaining predictors as *nuisance* variables, then we could add those control variables into the base model and measure the unique contribution(s) of adding the variable(s) of interest. In a more neutral or exploratory situation, we may want to evaluate the unique contribution of each variable when all other predictors are in the model. To do this for the intercept term (i.e., baseline reading ability), we can build a model that includes the effects of all four predictors on the intercept:

```
> eldel.intercepts <- lmer(pwmcor ~ Month +
                      wdspan1 + lk1 + ran1 + pa1 +
                      (Month | id),
                   data=ELDEL, REML=FALSE)
```

Then use the `drop1` function to evaluate the effects of removing (dropping) individual predictors from this model: `drop1` will cycle through fitting the relevant models and computing the comparisons and return the results in a compact form, we just need to specify the χ^2 test to make sure we get the right model comparison statistic:

```
> drop1(eldel.intercepts, test="Chisq")
Single term deletions

Model:
pwmcor ~ Month + wdspan1 + lk1 + ran1 + pa1 + (Month | id)
        Df  AIC   LRT  Pr(Chi)
<none>      5802
Month    1 6101 301.6 < 2e-16 ***
wdspan1  1 5800   0.2 0.69574
lk1      1 5814  14.5 0.00014 ***
ran1     1 5809   9.8 0.00172 **
pa1      1 5807   7.4 0.00670 **
---
Signif. codes:  0 '***' 0.001 '**' 0.01 '*' 0.05 '.' 0.1 ' ' 1
```

The `LRT` column is the likelihood ratio test statistic (i.e., the χ^2) and the right-most column is the corresponding *p*-value. It looks like the verbal span measure was not significantly related to baseline reading ability, but each of the other three variables uniquely contributed to the model.

Using the same strategy, we can evaluate the effect of these individual difference predictors on the rate of learning to read. We begin by fitting a full model that includes the effects of all predictors on the intercept and on the linear slope (`Month`):

```
> eldel.full <- lmer(pwmcor ~ Month *
                    (wdspan1 + lk1 + ran1 + pa1) +
                    (Month | id),
                  data=ELDEL, REML=FALSE)
```

then evaluate the effect of single term deletions (these will be just the effects on the slope):

```
> drop1(eldel.full, test="Chisq")
Single term deletions

Model:
pwmcor ~ Month * (wdspan1 + lk1 + ran1 + pa1) + (Month | id)
              Df  AIC   LRT  Pr(Chi)
<none>            5752
Month:wdspan1  1 5750  0.51 0.47405
Month:lk1      1 5763 13.24 0.00027 ***
```

```
Month:ran1    1 5758  8.08 0.00448 **
Month:pa1     1 5751  1.08 0.29779
---
Signif. codes:  0 '***' 0.001 '**' 0.01 '*' 0.05 '.' 0.1 ' ' 1
```

The results show that letter knowledge and rapid automatized naming significantly predicted the rate at which children learned to read (recall that letter knowledge and rapid automatized naming were only measured at baseline, so they were predictive in the temporal sense as well as the statistical sense). Phoneme awareness had been associated with reading ability at baseline, but did not predict the rate of learning to read.

Notice that we included the effect of `wdspan1` on the intercept in the full model even though the previous test had shown that this term did not significantly improve model fit. The effect of `wdspan1` on the intercept may not have been statistically significantly different from 0, but excluding it would have forced the model to treat it as if it were exactly 0, which could have a distorting effect on the estimation of the effect of `wdspan1` on the linear term (rate of learning to read). As a general rule, it is a bad idea to include higher-order effects without also including the lower-order effects. That said, since neither the intercept nor the slope effect of verbal span was significant, it would be reasonable to re-run the analyses excluding both terms from the model (in this case, the results were nearly identical).

7.2.1 Visualizing model fit

As demonstrated in previous chapters, we can plot the overall model fit by simply using the `fitted` function to get the model-fit values from the full model object. Note that we will need to explicitly exclude the missing `pwmcor` values – they wouldn't be plotted anyway (because they are missing) and they were excluded from the model fit, but if we don't exclude them, `ggplot` will return an error because the data sets will appear to be of different lengths. The resulting plot is shown in Figure 7.2.

```
> ggplot(subset(ELDEL, !is.na(pwmcor)), aes(Month, pwmcor)) +
    stat_summary(fun.y=mean, geom="point") +
    stat_summary(fun.data=mean_se, geom="errorbar", width=1) +
    stat_summary(aes(y=fitted(eldel.full)),
                 fun.y=mean, geom="line") +
    theme_bw(base_size=10) +
    labs(y="Picture-Word Matching Score")
```

Standard two-dimensional plots are great for representing two-variable data and additional categorical variables can be relatively easily represented by linetypes, point shapes, separate panels, etc. However, showing the relationships among more than two continuous variables can be very difficult. Heatmaps and contour plots can be effective for visualizing some three-variable

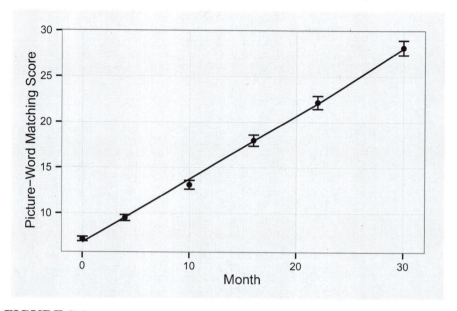

FIGURE 7.2
Overall development of reading ability (error bars indicate ±SE) with line showing the full model fit.

data sets. When other strategies fail, it may be necessary to discretize a variable — create a categorical variable out of a continuous one — in order to make an interpretable plot. Here is how we might do that in order to see the effect of letter knowledge on learning to read.

First, we need to create categorical variable that groups values of letter knowledge, for example, using a median split:

```
> ELDEL$LK <- factor(ELDEL$lk1 >= median(ELDEL$lk1),
                levels=c("FALSE", "TRUE"),
                labels=c("Low", "High"))
```

The `factor` function will turn its first input into a categorical variable with levels corresponding to the unique values in the set. In this case, that first input is a logical operation testing whether the letter knowledge score is greater than or equal to the median letter knowledge score. This will return logical values (`TRUE` or `FALSE`), which are not very informative level names, so the next two inputs tell the `factor` function to apply different labels to those levels. `FALSE` will become `Low` (values less than the median) and `TRUE` will become `High` (values greater than or equal to the median). A similar strategy could be used to divide letter knowledge into three, four, or more quantile groups by first using the `quantile` function to find the break points:

```
> b <- quantile(ELDEL$lk1[ELDEL$Month==0],
            probs=seq(0, 1, by=1/3))
```

then using the `cut` function to divide the letter knowledge variable at those break points and label the resulting factor levels:

```
> ELDEL$LK3 <- cut(ELDEL$lk1, breaks=b, include.lowest=TRUE,
                   labels=c("Low", "Medium", "High"))
```

Now we can plot the data using different shapes for the levels of letter knowledge and different linetypes for the model fits (Figure 7.3):

```
> ggplot(subset(ELDEL, !is.na(pwmcor)), aes(Month, pwmcor,
                                            shape=LK3)) +
    stat_summary(fun.y=mean, geom="point") +
    stat_summary(fun.data=mean_se, geom="errorbar", width=1) +
    stat_summary(aes(y=fitted(eldel.full), linetype=LK3),
                 fun.y=mean, geom="line") +
    theme_bw(base_size=10) +
    labs(y="Picture-Word Matching Score",
         shape="Letter\nKnowledge",
         linetype="Letter\nKnowledge") +
    theme(legend.position=c(0,1),
          legend.justification=c(0,1),
          legend.background=element_rect(color="black",
                                         fill="white"))
```

Note that the model used the continuous measure of letter knowledge, so this grouping is just a visualization aid that will apply in the same way to both the observed and model fit values.

7.3 Individual differences as random effects

Modeling individual differences as fixed effects is a good strategy when we have a measure of individual differences that is separate from the outcome variable itself, like a measure of letter knowledge that is separate from our measure of reading ability, and we can add that measure into the model. However, this is not always possible. For example, we might be interested in how effects in two conditions within the study are related to one another, or how between-participant differences in overall curve shape are related to within-participant differences between conditions (e.g., how processing speed differences are related to effect size). For these sorts of research questions, we can use the random effects to estimate effect sizes and use those effect sizes in subsequent analyses.

Let's consider an extremely simple case to see how random effects provide a way to quantify individual effect sizes. Figure 7.4 shows the performance Y of two participants (A and B) in two conditions (0 and 1). The dashed lines

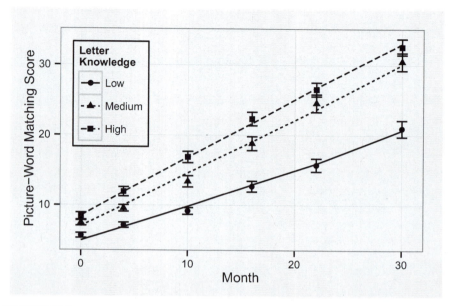

FIGURE 7.3
Development of reading ability (symbols, error bars indicate ±SE) grouped by tertile split on letter knowledge at study start. Lines indicate fit from the full growth model.

represent the condition means: $Y_{C=0} = 6, Y_{C=1} = 9$. The arrows represent the participant-by-condition random effects, each labeled with its corresponding ζ. We can use these random effects to compute effect sizes for individual subjects by computing the difference, for each participant, between the condition 1 random effect estimate and the condition 0 random effect estimate:

Participant A: $\zeta_{A1} - \zeta_{A0} = 1 - (-1) = 2$

Participant B: $\zeta_{B1} - \zeta_{B0} = (-1) - 1 = -2$

These effect sizes confirm the visually obvious fact that participant A had a larger effect size (difference between conditions) than participant B did. Indeed, in this minimal example, the difference between their effect sizes (4) is exactly the same value as we would get if we directly computed the difference between conditions for each participant (Participant A: 5; Participant B: 1) and took the difference between those scores. However, random effects are computed within a model of overall group performance, which offers a number of advantages that will be discussed in more detail in section 7.3.2.

Note that using random effects to compute individual effect sizes produces effect sizes that are symmetric around 0 – they correspond to the effect size *relative to the group mean effect size*. That is, participant B's -2 effect size does not necessarily indicate a reversal; rather, it means that the effect size

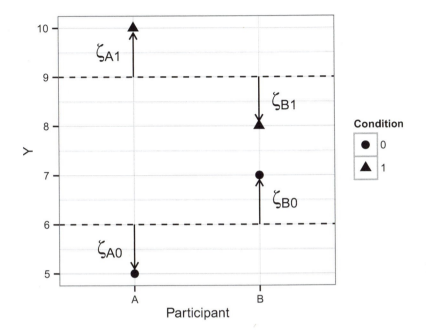

FIGURE 7.4
A simple example of two participants with different condition effect sizes. Dashed lines represent condition means; arrows represent random effect estimates.

was 2 units less than the overall group difference between conditions. We can think of this simple example as individual estimates of effect sizes based on the intercept term. In the context of a growth curve analysis, the same general approach can be used to quantify individual differences on linear, quadratic, etc., time course terms. The following example will demonstrate how to do this using data from an eye-tracking study.

7.3.1 Example: Function and thematic knowledge following stroke

These data come from an eye-tracking study of 17 participants with left hemisphere stroke (Kalénine, Mirman, & Buxbaum, 2012). Participants saw four pictures of objects and had to pick the one that matched the word they heard. The key research question was the time course of fixations to objects that were semantically related to the target compared to unrelated objects, and the semantic relationships could be either functional (e.g., *broom - sponge*, which serve a similar cleaning function) or thematic (e.g., *broom - dustpan*, which are used together). The overall time course is shown in Figure 7.5, though the more interesting question was whether there was any relationship between activation of these two different kinds of semantic knowledge across the 17 participants. Since the eye data *are* the measure of activation patterns, we don't have something to enter as a fixed effect into the model. We'll have to use the random effects to estimate effect sizes and then test correlations between those.

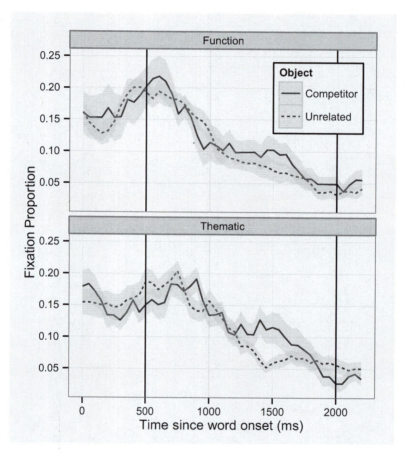

FIGURE 7.5

Distractor fixation proportion time course for function (top panel) and thematic (bottom panel) relations in 17 participants with left hemisphere stroke. Grey shading indicates ±SE; vertical black lines show the bounds of the analysis time window.

We will analyze data from 500ms to 2000ms after word onset and only the distractor fixations (not the target object fixations), so let's start by creating a GCA-appropriate subset data set:

```
> FunctThemePts.gca <- subset(FunctThemePts,
```

```
                                 Time >= 500 & Time <= 2000 &
                                 Object != "Target")
> summary(FunctThemePts.gca)
     subj            Condition           Object              Time
 206    : 124    Function:1054    Target    :   0    Min.    : 500
 281    : 124    Thematic:1054    Competitor:1054    1st Qu.: 850
 419    : 124                     Unrelated :1054    Median :1250
 2221   : 124                                        Mean   :1250
 1088   : 124                                        3rd Qu.:1650
 1238   : 124                                        Max.   :2000
 (Other):1364
    timeBin          meanFix           sumFix             N
 Min.   :30    Min.   :0.0000    Min.   : 0.00    Min.   :12.0
 1st Qu.:37    1st Qu.:0.0625    1st Qu.: 1.00    1st Qu.:15.0
 Median :45    Median :0.1000    Median : 2.00    Median :16.0
 Mean   :45    Mean   :0.1119    Mean   : 2.56    Mean   :15.4
 3rd Qu.:53    3rd Qu.:0.1667    3rd Qu.: 4.00    3rd Qu.:16.0
 Max.   :60    Max.   :0.5000    Max.   :12.00    Max.   :16.0
```

It will be easier to make and align the orthogonal polynomial time variable if we shift the `timeBin` variable so that the first bin in the analysis is labeled "1":

```
> FunctThemePts.gca$timeBin <- FunctThemePts.gca$timeBin - 29
```

Now we can create the fourth-order orthogonal polynomial:

```
> t <- poly((unique(FunctThemePts.gca$timeBin)), 4)
```

and insert it into the data frame:

```
> FunctThemePts.gca[, paste("ot", 1:4, sep="")] <-
                    t[FunctThemePts.gca$timeBin, 1:4]
```

The next step is to fit separate models for the distractor fixation time course in the function condition:

```
> m.funct <- lmer(meanFix ~ (ot1+ot2+ot3+ot4)*Object +
                           (ot1+ot2+ot3+ot4 | subj) +
                           (ot1+ot2 | subj:Object),
                  data=subset(FunctThemePts.gca,
                           Condition=="Function"),
                  control=lmerControl(optimizer="bobyqa"),
                  REML=FALSE)
```

and the thematic condition:

```
> m.theme <- lmer(meanFix ~ (ot1+ot2+ot3+ot4)*Object +
                           (ot1+ot2+ot3+ot4 | subj) +
                           (ot1+ot2 | subj:Object),
                  data=subset(FunctThemePts.gca,
                              Condition=="Thematic"),
                  control=lmerControl(optimizer="bobyqa"),
                  REML=FALSE)
```

Recall that the `ranef` function can be used to extract random effects. Our models will have two sets of random effects, which we can see using the `str` (structure) function (`vec.len` is an optional argument that determines how many of the "first few" elements are displayed for each vector; it is used here just to keep the output from running into the margins):

```
> str(ranef(m.funct), vec.len=2)
List of 2
 $ subj:Object:'data.frame':          34 obs. of  3 variables:
  ..$ (Intercept): num [1:34] -0.0327 0.0254 ...
  ..$ ot1        : num [1:34] 0.0339 0.0726 ...
  ..$ ot2        : num [1:34] -0.1538 0.0143 ...
 $ subj       :'data.frame':          17 obs. of  5 variables:
  ..$ (Intercept): num [1:17] 0.025 -0.0134 ...
  ..$ ot1        : num [1:17] 0.0924 -0.0454 ...
  ..$ ot2        : num [1:17] -0.1915 0.0375 ...
  ..$ ot3        : num [1:17] 0.1124 -0.0677 ...
  ..$ ot4        : num [1:17] 0.0386 0.0474 ...
 - attr(*, "class")= chr "ranef.mer"
```

the `head` function is also useful for seeing how a data frame is set up:

```
> head(ranef(m.funct)$"subj:Object")
                 (Intercept)        ot1         ot2
206:Competitor  -0.03274632  0.033860 -0.1538105
206:Unrelated    0.02538758  0.072615  0.0143118
281:Competitor   0.00664979 -0.043948 -0.0596063
281:Unrelated   -0.02003123 -0.055298  0.1158190
419:Competitor  -0.00063844  0.246673 -0.0012456
419:Unrelated   -0.00929344 -0.152940  0.0568354
```

The participant:object codes are row names, but it will be helpful to make them variables in the data frame and divide the participant and object pieces into separate variables. We can get the row names using the `rownames` function and split them up using the `colsplit` function (from the `reshape2` package):

```
> re.id <- colsplit(row.names(ranef(m.funct)$"subj:Object"),
                    ":", c("Subject", "Object"))
```

Now we combine them with the random effect estimates:

```
> re.funct <- data.frame(re.id, ranef(m.funct)$"subj:Object")
> head(re.funct)
                 Subject       Object X.Intercept.        ot1
206:Competitor       206   Competitor  -0.03274632   0.033860
206:Unrelated        206    Unrelated   0.02538758   0.072615
281:Competitor       281   Competitor   0.00664979  -0.043948
281:Unrelated        281    Unrelated  -0.02003123  -0.055298
419:Competitor       419   Competitor  -0.00063844   0.246673
419:Unrelated        419    Unrelated  -0.00929344  -0.152940
                          ot2
206:Competitor     -0.1538105
206:Unrelated       0.0143118
281:Competitor     -0.0596063
281:Unrelated       0.1158190
419:Competitor     -0.0012456
419:Unrelated       0.0568354
```

To compute the effect size, we need to compute the difference between the competitor and unrelated random effect estimates for each participant. The `ddply` function from the `plyr` package provides a simple way to do this (this function was introduced in Chapter 4):

```
> ES.funct <- ddply(re.funct, .(Subject), summarize,
    Function_Intercept = X.Intercept.[Object=="Competitor"] -
                         X.Intercept.[Object=="Unrelated"],
    Function_Linear = ot1[Object=="Competitor"] -
                      ot1[Object=="Unrelated"])
```

Now we do the same steps for the thematic condition, first extracting the random effects:

```
> re.theme <- data.frame(
              colsplit(row.names(ranef(m.theme)$"subj:Object"),
                    ":", c("Subject", "Object")),
              ranef(m.theme)$"subj:Object")
```

then computing the effect sizes:

```
> ES.theme <- ddply(re.theme, .(Subject), summarize,
    Thematic_Intercept = X.Intercept.[Object=="Competitor"] -
                         X.Intercept.[Object=="Unrelated"],
    Thematic_Linear = ot1[Object=="Competitor"] -
                      ot1[Object=="Unrelated"])
```

Finally, we combine the function and thematic condition effect sizes into a single data frame:

```
> ES <- merge(ES.funct, ES.theme)
> head(ES)
  Subject Function_Intercept Function_Linear Thematic_Intercept
1    206          -0.0581339       -0.038755          0.0309626
2    281           0.0266810        0.011349          0.0153672
3    419           0.0086550        0.399613         -0.0018655
4   1088          -0.0032830       -0.156276         -0.0845982
5   1238          -0.0133492       -0.139859         -0.0220510
6   1392          -0.0031964        0.191226          0.0615263
  Thematic_Linear
1      -0.1528792
2       0.0039936
3      -0.1459968
4      -0.0619093
5      -0.0155540
6      -0.3531367
```

This data frame contains all of the data we need to test the correlation between function and thematic condition effect sizes. Starting with the intercept term (i.e., overall fixation proportion for the related competitors compared to the unrelated distractors):

```
> cor.test(ES$Function_Intercept, ES$Thematic_Intercept)
        Pearson's product-moment correlation

data:  ES$Function_Intercept and ES$Thematic_Intercept
t = -2.3602, df = 15, p-value = 0.03223
alternative hypothesis: true correlation is not equal to 0
95 percent confidence interval:
 -0.80075 -0.05300
sample estimates:
     cor
-0.52039
```

For the intercept term, there was a significant negative correlation ($r = -0.52$, $p = 0.032$), indicating that participants who showed larger function competition effects tended to show smaller thematic competition effects, and vice versa. There was a similar negative correlation pattern for the linear term ($r = -0.65$, $p = 0.0043$).

```
> cor.test(ES$Function_Linear, ES$Thematic_Linear)
        Pearson's product-moment correlation

data:  ES$Function_Linear and ES$Thematic_Linear
t = -3.3571, df = 15, p-value = 0.004322
alternative hypothesis: true correlation is not equal to 0
```

```
95 percent confidence interval:
 -0.86372 -0.25445
sample estimates:
     cor
-0.65499
```

Figure 7.6 shows these correlations as scatterplots of the individual participants' function and thematic competition effect sizes on the intercept and linear terms.

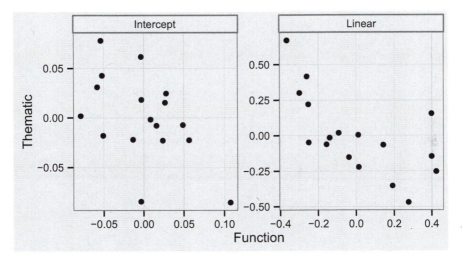

FIGURE 7.6
Scatterplots of individual function and thematic competition effect sizes on intercept (left) and linear (right) terms.

7.3.2 Individual differences, not individual models

The simple example in Figure 7.4 might suggest a simpler alternative: why not just fit separate models for each participant and examine the individual effect sizes estimated by individual participant models? The problem with this approach is that it will *over*-estimate the individual differences by considering each participant independently of the group. By modeling the group data and the individual variation within a single model, multilevel (growth curve) models provide an estimate of individual effect sizes that combine each individual's pattern with the overall pattern of the group – of which that individual is (presumably) a member. This is closely related to the shrinkage issue discussed in Chapter 4: effect size estimates from multilevel models will tend to be pulled (shrunk) toward the overall group mean.

We can demonstrate this in a simple example with simulated reaction time (RT) data for 15 participants in two conditions. For the "easy" condition, the reaction times were drawn from a normal distribution with a mean of 800ms and a standard deviation of 30ms. The individual condition effect sizes were drawn from a normal distribution with a mean of 20ms (i.e., performance in the "hard" condition was about 20ms slower) and a standard deviation of 10ms. A summary of the generated data is shown in Table 7.1.

TABLE 7.1

Simulated RT Data: Mean (SD) for Easy and Hard Conditions, Difference between Means and Results of a Paired-Samples t-Test.

Easy	Hard	Difference	t	p
801 (25.9)	822 (28.4)	21.2	7.9	1.52e-06

Using these simulated data, we can compute individual effect sizes in two different ways: (1) using random effect estimates from a multilevel model of the full group data and (2) by individually subtracting the easy condition RT from the hard condition RT for each participant (analogous to fitting individual models for each participant). The average of the estimated effect sizes is the same[1] for both approaches (corresponding to the difference value in Table 7.1), but the individual-models approach produces a much wider distribution of individual effect sizes, as shown in the left panel of Figure 7.7. In other words, the multilevel model partitions the variance among overall group effects, individual participant effects, and noise, whereas the individual models assign all variability to individual participant effects.

The same pattern arises if we re-analyze the eye-tracking data from section 7.3.1 using separate fourth-order orthogonal polynomial models for each individual participant. The right panel in Figure 7.7 shows the individual participant effect sizes on the linear term in the function condition estimated from a group model's random effects as in section 7.3.1 and from individual participant models. As in the simulated data example, fitting independent models for each participant produces a wider distribuction of effect size estimates. In a sense, the problem with using individual models is the reverse of the problem with using t-tests and ANOVAs: whereas t-tests and ANOVAs treat all variability as noise, individual models treat all variability as individual differences. Multilevel models describe the data simultaneously at the group and individual levels, so they partition the variability among group effects (fixed effects), participant effects (random effects), and residual error. As a result, their individual effect size estimates take into account an estimate of the overall group-level effect size and truly random noise to arrive at a more informed estimate of individual differences. (For more discussion of these issues and another real data example see Kliegl et al., 2011; for an accessible statistical discussion of shrinkage and individual estimates see Efron & Morris, 1977).

---

[1]For the group model effect sizes, the fixed effect (group average) was added to the random effect estimates to get true effect size estimates rather than relative effect size estimates.

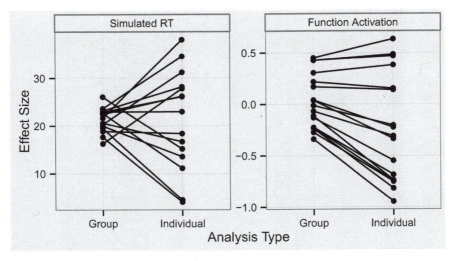

FIGURE 7.7
Comparisons of individual effect size estimates based on random effects from a group model and based on individual participant models. Left: Simulated RT data. Right: Function activation linear term.

7.3.3   Using effect sizes for subsequent analyses

Another context for using random effects to quantify individual effect sizes is when you need to quantify individual differences for an analysis that can't practically be implemented within GCA. For example, voxel-based lesion-symptom mapping (e.g., Bates et al., 2003; Schwartz et al., 2011) is a technique for investigating the neural basis of behavior. The general approach is to test each small region of brain tissue (voxel) for behavioral differences between participants with vs. without damage in that region (a correction for multiple comparisons is typically applied). The individual tests are usually t-tests or simple regressions, which are easy to implement and fast to execute. Adapting this method to GCA would mean, for each voxel, fitting a separate growth curve model with brain damage status of that voxel as a fixed effect. Even if software were developed to make this possible, it would be impractical because it would require fitting many thousands of growth curve models (possibly hundreds of thousands), which could take weeks to compute even if each model only takes a few seconds. A more practical approach is to compute individual participant effect size estimates as described in this chapter and use them as the behavioral measure in a standard VLSM analysis. More generally, using random effects to compute individual effect sizes makes them available for analyses that are impossible or impractical to integrate into GCA.

7.4   Chapter recap

This chapter described and demonstrated two methods for using growth curve analysis (and multilevel regression more generally) to analyze individual differences. The first method is simply an extension of multiple regression to the GCA framework: add the individual difference measure as a fixed effect into the growth curve model. This allows testing the effects of individual differences on different aspects of the growth curves (intercept, linear slope, quadratic, etc.). The standard properties and concerns of multiple regression also apply to this method (continuous vs. categorical predictors, collinear predictors, etc.).

The second method is to use the random effects to estimate individual participant effect sizes. This approach is useful when there is no separate measure that could be added as a fixed effect into the model, or doing so would be computationally impractical. This method makes use of the fact that multilevel models estimate group-level (fixed) effects, individual-level (random) effects, and residual error. The individual-level random effects provide estimates of how individuals deviate from the overall group pattern, which is a measure of relative effect size that takes into consideration the overall group pattern as well as the residual error. In essence, the model's estimate of the group-level effect serves as an anchor that allows the model to estimate individual effect sizes as corresponding to individual deviation from the group mean and residual noise. This tends to shrink estimates of individual effect sizes toward the group mean compared to fitting separate models for each individual participant, which cannot estimate the group-level effect and therefore attributes all variability to individual differences. On the assumption that the participants in the sample do constitute a meaningful group with an informative group-level effect, fitting separate models for individual participants tends to over-estimate the variability of individual effect sizes.

7.5   Exercises

The `CohortRhyme` example data set contains data from an eye-tracking experiment (Mirman et al., 2011) similar to the one described in section 7.3.1, except this one investigated phonological competition between *cohorts* (e.g., *penny–pencil*) and *rhymes* (e.g., *carrot–parrot*). Three groups of participants were tested: five individuals with Broca's aphasia, three individuals with Wernicke's aphasia, and 12 control participants.

1. Use fourth-order orthogonal polynomials to analyze (separately) the cohort and rhyme competition effects.

(a) Test group differences in cohort and rhyme competition effects.

(b) Evaluate all pairwise group comparisons for different time terms (see Chapter 5).

(c) Make a multi-panel plot that shows cohort and rhyme competition effects for each of the groups (you may want to use `facet_grid`).

2. Compute individual participant's cohort and rhyme competition effects on the intercept, linear, and quadratic time terms (tip: remember to remove the group fixed effect so that these effect sizes will be relative to the overall mean, not the diagnosis group mean).

3. Test correlations between cohort and rhyme effect sizes for the full set of participants and separately for the control and aphasic participants.

4. Make a multi-panel scatterplot that shows the correlations for the intercept, linear, and quadratic terms in separate panels as in Figure 7.6 (tip: you may need to use `dcast`, which is the companion function to `melt` that lets you convert from a long data format to a wide data format).

8

Complete examples

CONTENTS

The purpose of this chapter is to provide simple and complete templates for analysis code, model output, and example write-ups. These are meant to serve as a "quick reference" for looking up code syntax without wading through the narrative of the main book text. Each example will begin with a quick reminder of the data, then provide the code for analysis with minimal comments so you can copy and paste it. The # can be used to mark comments in R code – these lines are not executed by R and it is good programming practice to use comments so that you can have a record of what each section of code is supposed to do. For detailed discussion of each example, refer back to the appropriate chapter for each example. The code blocks will include model output and code for plotting the observed data and model fits. Each example will conclude with a brief write-up demonstrating how the analysis strategy and results could be reported in a journal article or other scientific publication.

For each of the examples, the `lme4` and **ggplot2** packages will need to be loaded:

```
> library(lme4)
> library(ggplot2)
```

8.1 Linear change

These data are from a randomized placebo-controlled study of the effect of amantadine on recovery from brain injury (Giacino et al., 2012). After a baseline assessment, patients who were in a vegetative or minimally conscious state received either amantadine or placebo for 4 weeks and their functional recovery was measured using the Disability Rating Scale (DRS).

```
> # inspect the data to check variable names and types
> summary(amant.ex)
    Patient            Group          Week           DRS
  1008   :  5    Placebo    :85   Min.   :0    Min.    : 7.0
  1009   :  5    Amantadine:65   1st Qu.:1    1st Qu.:17.0
  1017   :  5                     Median :2    Median :20.5
  1042   :  5                     Mean   :2    Mean    :19.3
  1044   :  5                     3rd Qu.:3    3rd Qu.:22.0
  1054   :  5                     Max.   :4    Max.    :28.0
  (Other):120
> # fit base model
> amant.base <- lmer(DRS ~ 1 + Week + (1 + Week | Patient),
                    data=amant.ex, REML=F)
> # gradually add effects of treatment group
> # add effect of group on intercept (baseline assessment)
> amant.0 <- lmer(DRS ~ 1 + Week+Group + (1 + Week | Patient),
                  data=amant.ex, REML=F)
> # add effect of group on linear slope (rate of recovery)
> amant.1 <- lmer(DRS ~ 1 + Week*Group + (1 + Week | Patient),
                  data=amant.ex, REML=F)
> # compare models
> anova(amant.base, amant.0, amant.1)
Data: amant.ex
Models:
amant.base: DRS ~ 1 + Week + (1 + Week | Patient)
amant.0: DRS ~ 1 + Week + Group + (1 + Week | Patient)
amant.1: DRS ~ 1 + Week * Group + (1 + Week | Patient)
           Df AIC BIC logLik deviance Chisq Chi Df Pr(>Chisq)
amant.base  6 622 641   -305      610
amant.0     7 623 644   -304      609  1.63      1     0.202
amant.1     8 619 643   -302      603  5.56      1     0.018 *
---
Signif. codes:  0 '***' 0.001 '**' 0.01 '*' 0.05 '.' 0.1 ' ' 1
> # examine parameter estimates from full model
> coef(summary(amant.1))
                       Estimate Std. Error t value
(Intercept)            22.05882    0.48485 45.4964
Week                   -0.70000    0.22117 -3.1650
GroupAmantadine        -1.42805    0.73654 -1.9389
Week:GroupAmantadine   -0.83077    0.33598 -2.4726
```

```
> # plot data with full model fit
> p1 <- ggplot(amant.ex, aes(Week, DRS, shape=Group)) +
        stat_summary(fun.data=mean_se, geom="pointrange") +
        stat_summary(aes(y=fitted(amant.1), linetype=Group),
                     fun.y=mean, geom="line") +
        theme_bw(base_size=10)
> # save this plot as a PDF file, specify size and resolution
> ggsave("amantFit.pdf", p1, width=4.5, height=3.5, dpi=300)
```

Example write-up:

> The data were fit using a linear growth model with fixed effects
> of group (amantadine vs. placebo) on the intercept (baseline as-
> sessment) and linear (rate of recovery) terms and random effects
> of participants on the intercept and slope to model individual dif-
> ferences in initial severity and rate of recovery. The fixed effects of
> group were added individually and their effects on model fit were
> evaluated using model comparisons. Improvements in model fit
> were evaluated using -2 times the change in log-likelihood, which
> is distributed as χ^2 with degrees of freedom equal to the number
> of parameters added. All analyses were carried out in R version
> 3.0.2 using the lme4 package (version 1.0-5).

> The data and model fits are shown in Figure 8.1. There was no ef-
> fect of group on the intercept ($\chi^2(1) = 1.63, p = 0.202$), indicating
> that there were no baseline differences between the groups (as ex-
> pected, because participants were randomly assigned to groups).
> There was a significant effect of group on the linear slope term
> ($\chi^2(1) = 5.56, p = 0.0183$), indicating faster recovery in the aman-
> tadine group compared to the placebo group. This difference was
> approximately 0.831 ($SE = 0.336$) points per week faster recovery
> for the amantadine group compared to the placebo group.

8.2 Orthogonal polynomials

These data are from an experiment testing the effect of transitional probability
(TP) on learning novel words (Mirman, Magnuson, et al., 2008). Participants
learned a set of novel words through trial-and-error: on each trial they saw two
novel objects and heard a novel word, picked which object had been named,
and received feedback. The outcome variable is accuracy (proportion correct
responses), which starts out near chance (approx. 50% correct) and gradually
reaches an asymptote around 90% correct.

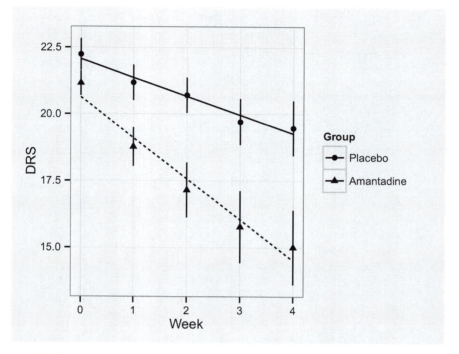

FIGURE 8.1
Observed data (symbols, vertical lines indicate ±SE) and linear model fits (lines) for functional recovery for the placebo and amantadine groups.

```
> # inspect the data to check variable names and types
> summary(WordLearnEx)

    Subject          TP           Block           Accuracy
   244    : 10   Low :280   Min.   : 1.0   Min.   :0.000
   253    : 10   High:280   1st Qu.: 3.0   1st Qu.:0.667
   302    : 10              Median : 5.5   Median :0.833
   303    : 10              Mean   : 5.5   Mean   :0.805
   305    : 10              3rd Qu.: 8.0   3rd Qu.:1.000
   306    : 10              Max.   :10.0   Max.   :1.000
   (Other):500

> # create second-order orthogonal polynomial
> t <- poly(unique(WordLearnEx$Block), 2)
> # create orthogonal polynomial time variables in data frame
> WordLearnEx[,paste("ot", 1:2, sep="")] <-
                         t[WordLearnEx$Block, 1:2]
> # fit base model
> WordLearn.base <- lmer(Accuracy ~ (ot1+ot2) +
                         (ot1+ot2 | Subject),
```

```
                              data=WordLearnEx, REML=FALSE)
> # add effect of TP on intercept
> WordLearn.0 <- lmer(Accuracy ~ (ot1+ot2) + TP +
                              (ot1+ot2 | Subject),
                    data=WordLearnEx, REML=FALSE)
> # add effect of TP on linear term
> WordLearn.1 <- lmer(Accuracy ~ (ot1+ot2) + TP + ot1:TP +
                              (ot1+ot2 | Subject),
                    data=WordLearnEx, REML=FALSE)
> # add effect of TP on quadratic term
> WordLearn.2 <- lmer(Accuracy ~ (ot1+ot2)*TP +
                              (ot1+ot2 | Subject),
                    data=WordLearnEx, REML=FALSE)


> # compare models
> anova(WordLearn.base, WordLearn.0,
        WordLearn.1, WordLearn.2)


Data: WordLearnEx
Models:
WordLearn.base: Accuracy ~ (ot1 + ot2) + (ot1 + ot2 |
Subject)
WordLearn.0: Accuracy ~ (ot1 + ot2) + TP + (ot1 + ot2 |
Subject)
WordLearn.1: Accuracy ~ (ot1 + ot2) + TP + ot1:TP + (ot1 +
ot2 | Subject)
WordLearn.2: Accuracy ~ (ot1 + ot2) * TP + (ot1 + ot2 |
Subject)
Df AIC BIC logLik deviance Chisq Chi Df Pr(>Chisq)
WordLearn.base 10 -331 -288 175 -351
WordLearn.0 11 -330 -283 176 -352 1.55 1 0.213
WordLearn.1 12 -329 -277 176 -353 0.36 1 0.550
WordLearn.2 13 -333 -276 179 -359 5.95 1 0.015

WordLearn.base
WordLearn.0
WordLearn.1
WordLearn.2 *
---
Signif. codes: 0 '***' 0.001 '**' 0.01 '*' 0.05 '.' 0.1 ' '
1


> # get parameter estimates and estimate p-values
> WordLearn.coefs <- data.frame(coef(summary(WordLearn.2)))
```

```
> WordLearn.coefs$p <-
                2*(1-pnorm(abs(WordLearn.coefs$t.value)))
> WordLearn.coefs
                Estimate Std..Error    t.value              p
(Intercept)   0.7785250    0.021728  35.830648  0.0000e+00
ot1           0.2863155    0.037789   7.576772  3.5527e-14
ot2          -0.0508493    0.033188  -1.532182  1.2548e-01
TPHigh        0.0529607    0.030728   1.723538  8.4791e-02
ot1:TPHigh    0.0010754    0.053441   0.020123  9.8395e-01
ot2:TPHigh   -0.1164548    0.046934  -2.481234  1.3093e-02
> # plot data with full model fit
> p2 <- ggplot(WordLearnEx, aes(Block, Accuracy, shape=TP)) +
    stat_summary(aes(y=fitted(WordLearn.2), linetype=TP),
                fun.y=mean, geom="line") +
    stat_summary(fun.data=mean_se, geom="pointrange") +
    theme_bw(base_size=10) +
    coord_cartesian(ylim=c(0.5, 1.0)) +
    scale_shape_manual(values=c(1,16)) +
    scale_x_continuous(breaks=1:10)
> # save as a PDF file
> ggsave("WordLearnExFit.pdf", p2,
        width=4.5, height=4, dpi=300)
```

Example write-up:

Growth curve analysis (Mirman, 2014) was used to analyze the learning of the novel words over the course of 10 training blocks. The overall learning curves were modeled with second-order orthogonal polynomials and fixed effects of TP on all time terms. The low TP condition was treated as the baseline and parameters were estimated for the high TP condition. The model also included random effects of participants on all time terms. The fixed effects of TP were added individually and their effects on model fit were evaluated using model comparisons. Parameter-specific p-values were estimated using the normal approximation (i.e., treating the t-value as a z-value). All analyses were carried out in R version 3.0.2 using the lme4 package (version 1.0-5).

The effect of TP on the intercept did not improve model fit $(\chi^2(1) = 1.55, p = 0.213)$, nor did the effect of TP on the linear term $(\chi^2(1) = 0.358, p = 0.55)$. The effect of TP on the quadratic term, however, did improve model fit $(\chi^2(1) = 5.95, p = 0.0147)$, indicating that the low and high TP conditions differed in the rate of word learning. The data and model fits are shown in Figure 8.2 and Table 8.1 shows the fixed effect parameter estimates and their standard errors along with corresponding t- and p-values.

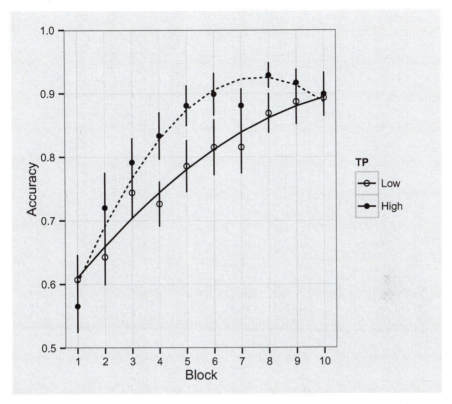

FIGURE 8.2
Observed data (symbols, vertical lines indicate ±SE) and growth curve model fits (lines) for effect of transitional probability (TP) on novel word learning.

TABLE 8.1
Parameter Estimates for Analysis of Effect of TP on Novel Word Learning

	Estimate	Std. Error	t	p
Intercept	0.779	0.022	35.831	0.000
Linear	0.286	0.038	7.577	0.000
Quadratic	-0.051	0.033	-1.532	0.125
High TP: Intercept	0.053	0.031	1.724	0.085
High TP: Linear	0.001	0.053	0.020	0.984
High TP: Quadratic	-0.116	0.047	-2.481	0.013

8.3 Within-subject manipulation

These data come from an eye-tracking study of how word frequency affects the time course of spoken word recognition. Participants were presented with four pictures of familiar objects, heard a spoken word, and had to click on the matching object. Some of words were high frequency (very common) and some were low frequency (relatively less common); each participant heard both kinds of words. Participants' eye movements were tracked and the critical outcome measure is proportion of fixations on the target object in successive 50ms time bins.

```
> # inspect the data to check variable names and types
> summary(TargetFix)
     Subject         Time           timeBin    Condition
 708    : 30    Min.   : 300    Min.   : 1    High:150
 712    : 30    1st Qu.: 450    1st Qu.: 4    Low :150
 715    : 30    Median : 650    Median : 8
 720    : 30    Mean   : 650    Mean   : 8
 722    : 30    3rd Qu.: 850    3rd Qu.:12
 725    : 30    Max.   :1000    Max.   :15
 (Other):120
     meanFix              sumFix              N
 Min.   :0.0286    Min.   : 1.0    Min.   :33.0
 1st Qu.:0.2778    1st Qu.:10.0    1st Qu.:35.8
 Median :0.4558    Median :16.0    Median :36.0
 Mean   :0.4483    Mean   :15.9    Mean   :35.5
 3rd Qu.:0.6111    3rd Qu.:21.2    3rd Qu.:36.0
 Max.   :0.8286    Max.   :29.0    Max.   :36.0
> # create third-order orthogonal polynomial
> t <- poly(unique(TargetFix$timeBin), 3)
> # create orthogonal polynomial time variables in data frame
> TargetFix[,paste("ot", 1:3, sep="")] <-
                            t[TargetFix$timeBin, 1:3]
> # fit full model
> TargFix.full <- lmer(meanFix ~ (ot1+ot2+ot3)*Condition +
                       (ot1+ot2+ot3 | Subject) +
                       (ot1+ot2+ot3 |
                          Subject:Condition),
                  control=lmerControl(optimizer="bobyqa"),
                  data=TargetFix, REML=FALSE)
> # get parameter estimates and estimate p-values
> TargFix.coefs <- data.frame(coef(summary(TargFix.full)))
> TargFix.coefs$p <-
              2*(1-pnorm(abs(TargFix.coefs$t.value)))
> TargFix.coefs
                Estimate Std..Error    t.value          p
(Intercept)   0.47732275   0.013852  34.4577764  0.0000e+00
ot1           0.63856037   0.059935  10.6541796  0.0000e+00
ot2          -0.10959793   0.038488  -2.8475730  4.4054e-03
```

ot3	-0.09326119	0.023302	-4.0022000	6.2756e-05
ConditionLow	-0.05811224	0.018787	-3.0932260	1.9799e-03
ot1:ConditionLow	0.00031882	0.065786	0.0048463	9.9613e-01
ot2:ConditionLow	0.16354551	0.053930	3.0325447	2.4250e-03
ot3:ConditionLow	-0.00208691	0.027044	-0.0771678	9.3849e-01

```
> # plot data with full model fit
> p3 <- ggplot(TargetFix,
               aes(Time, meanFix, shape=Condition)) +
    stat_summary(fun.y=mean, geom="point") +
    stat_summary(fun.data=mean_se, geom="errorbar") +
    stat_summary(aes(y=fitted(TargFix.full),
                 linetype=Condition),
                 fun.y=mean, geom="line") +
    theme_bw(base_size=10) +
    labs(x="Time since word onset (ms)",
         y="Fixation proportion",
         shape="Word\nFrequency", linetype="Word\nFrequency")
> # save plot as a PDF file
> ggsave("TargFixFit.pdf", p3, width=4.5, height=3, dpi=300)
```

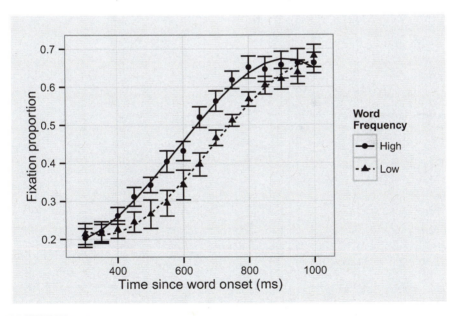

FIGURE 8.3
Observed data (symbols, error bars indicate ±SE) and growth curve model fits (lines) for effect of word frequency on the time course of spoken word recognition.

Example write-up:

Growth curve analysis (Mirman, 2014) was used to analyze the time course of fixation from 300ms to 1000ms after word onset (i.e., from the earliest word-driven fixations to when target fixations had plateaued). The overall time course of target fixations was captured with a third-order (cubic) orthogonal polynomial with fixed effects of condition (low vs. high frequency) on all time terms, and participant and participant-by-condition random effects on all time terms. The high frequency condition was treated as the reference (baseline) and relative parameters estimated for the low frequency condition. Statistical significance (*p*-values) for individual parameter estimates was assessed using the normal approximation (i.e., treating the *t*-value as a *z*-value). All analyses were carried out in R version 3.0.2 using the lme4 package (version 1.0-5).

The data and model fits are shown in Figure 8.3. There was a significant effect of frequency on the intercept ($Estimate = -0.0581$, $SE = 0.0188$, $p = 0.00198$), indicating overall higher fixation probability for the high frequency words than the low frequency words. There was also an effect of frequency on the quadratic term ($Estimate = 0.164$, $SE = 0.0539$, $p = 0.00243$), reflecting faster recognition of high frequency words than low frequency words. Frequency did not have significant effects on the linear or cubic terms (both $p > 0.9$).

8.4   Logistic GCA

These are the same data as in the previous example: the time course of target picture fixation from a study in which participants had to pick which of four pictures matched a spoken word. For this example, the outcome variable is a pair of integer values: the number of target fixations and the number of non-target fixations. Construction of the third-order orthogonal polynomial will not be repeated here because it is identical to the previous example.

```
> # inspect the data to check variable names and types
> summary(TargetFix)
    Subject         Time          timeBin    Condition
 708    : 30   Min.   : 300   Min.   : 1   High:150
 712    : 30   1st Qu.: 450   1st Qu.: 4   Low :150
 715    : 30   Median : 650   Median : 8
 720    : 30   Mean   : 650   Mean   : 8
 722    : 30   3rd Qu.: 850   3rd Qu.:12
 725    : 30   Max.   :1000   Max.   :15
```

```
(Other):120
     meanFix              sumFix               N                  ot1
 Min.   :0.0286    Min.    : 1.0     Min.    :33.0     Min.    :-0.418
 1st Qu.:0.2778    1st Qu.:10.0      1st Qu.:35.8      1st Qu.:-0.239
 Median :0.4558    Median :16.0      Median :36.0      Median : 0.000
 Mean   :0.4483    Mean    :15.9     Mean    :35.5     Mean    : 0.000
 3rd Qu.:0.6111    3rd Qu.:21.2      3rd Qu.:36.0      3rd Qu.: 0.239
 Max.   :0.8286    Max.    :29.0     Max.    :36.0     Max.    : 0.418

       ot2                 ot3
 Min.   :-0.2906    Min.    :-0.456
 1st Qu.:-0.2283    1st Qu.:-0.246
 Median :-0.0415    Median : 0.000
 Mean    : 0.0000   Mean    : 0.000
 3rd Qu.: 0.2699    3rd Qu.: 0.246
 Max.    : 0.4723   Max.    : 0.456
> # fit full model
> # Note: The full random effects model did not converge,
> #    so the random effects were simplified by removing
> #    the cubic term from the Subject:Condition random effect
> TargFix.log <- glmer(cbind(sumFix, N-sumFix) ~
                       (ot1+ot2+ot3)*Condition +
                       (ot1+ot2+ot3 | Subject) +
                       (ot1+ot2 | Subject:Condition),
                     data=TargetFix, family=binomial)
> # inspect parameter estimates
> coef(summary(TargFix.log))
                  Estimate Std. Error  z value    Pr(>|z|)
(Intercept)      -0.116808   0.065421 -1.78549  7.4182e-02
ot1               2.818568   0.298037  9.45709  3.1663e-21
ot2              -0.558929   0.169087 -3.30558  9.4781e-04
ot3              -0.320873   0.127327 -2.52006  1.1733e-02
ConditionLow     -0.261509   0.090896 -2.87701  4.0146e-03
ot1:ConditionLow  0.064199   0.330985  0.19396  8.4621e-01
ot2:ConditionLow  0.695116   0.239426  2.90327  3.6929e-03
ot3:ConditionLow -0.070570   0.165858 -0.42548  6.7049e-01
> # plot data with full model fit
> p4 <- ggplot(TargetFix, aes(Time,meanFix,shape=Condition))+
    stat_summary(fun.y=mean, geom="point") +
    stat_summary(fun.data=mean_se, geom="errorbar") +
    stat_summary(aes(y=fitted(TargFix.log), linetype=Condition),
                 fun.y=mean, geom="line") +
    theme_bw(base_size=10) +
    labs(x="Time since word onset (ms)",
        y="Fixation proportion",
```

```
              shape="Word\nFrequency", linetype="Word\nFrequency")
> # save plot as a PDF file
> ggsave("TargFixLogisticFit.pdf", p4,
          width=4.5, height=3, dpi=300)
```

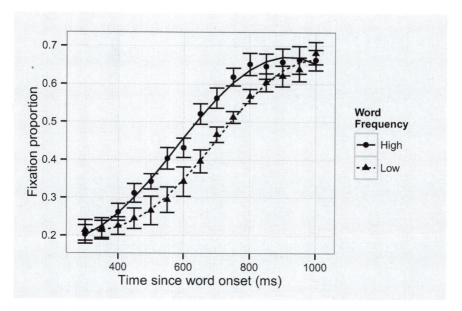

FIGURE 8.4

Observed target fixation proportion (symbols, error bars indicate ±SE) and logistic growth curve model fits (lines) for effect of word frequency on the time course of spoken word recognition.

Example write-up:

> Logistic growth curve analysis (Mirman, 2014) was used to analyze the time course of fixation from 300ms to 1000ms after word onset (i.e., from the earliest word-driven fixations to when target fixations had plateaued). The overall time course of target fixations was captured with a third-order (cubic) orthogonal polynomial with fixed effects of condition (low vs. high frequency) on all time terms, and participant and participant-by-condition random effects on all time terms except the cubic (the model did not converge with the full random effect structure, so this higher-order term was removed because the cubic term was not expected to capture key frequency condition differences). The high frequency condition was treated as the reference (baseline) and relative parameters estimated for the low frequency condition. Statistical significance (p-values) for individual parameter estimates was assessed using

the normal approximation (i.e., treating the *t*-value as a *z*-value). All analyses were carried out in R version 3.0.2 using the lme4 package (version 1.0-5).

The data and model fits are shown in Figure 8.4. There was a significant effect of frequency on the intercept ($Estimate = -0.262$, $SE = 0.0909$, $p = 0.00401$), indicating overall higher odds of fixating the target for the high frequency words than the low frequency words. There was also an effect of frequency on the quadratic term ($Estimate = 0.695$, $SE = 0.239$, $p = 0.00369$), reflecting faster recognition of high frequency words than low frequency words. Word frequency did not have significant effects on the linear or cubic terms (both $p > 0.6$).

8.5 Quasi-logistic GCA

Continuing with the data from the previous two examples, this example will use the *empirical logit* transformation to approximate logistic regression.

```
> # inspect the data to check variable names and types
> summary(TargetFix)
    Subject         Time          timeBin    Condition
 708    : 30   Min.   : 300   Min.   : 1   High:150
 712    : 30   1st Qu.: 450   1st Qu.: 4   Low :150
 715    : 30   Median : 650   Median : 8
 720    : 30   Mean   : 650   Mean   : 8
 722    : 30   3rd Qu.: 850   3rd Qu.:12
 725    : 30   Max.   :1000   Max.   :15
 (Other):120
    meanFix           sumFix            N              ot1
 Min.   :0.0286   Min.   : 1.0   Min.   :33.0   Min.   :-0.418
 1st Qu.:0.2778   1st Qu.:10.0   1st Qu.:35.8   1st Qu.:-0.239
 Median :0.4558   Median :16.0   Median :36.0   Median : 0.000
 Mean   :0.4483   Mean   :15.9   Mean   :35.5   Mean   : 0.000
 3rd Qu.:0.6111   3rd Qu.:21.2   3rd Qu.:36.0   3rd Qu.: 0.239
 Max.   :0.8286   Max.   :29.0   Max.   :36.0   Max.   : 0.418

      ot2              ot3
 Min.   :-0.2906   Min.   :-0.456
 1st Qu.:-0.2283   1st Qu.:-0.246
 Median :-0.0415   Median : 0.000
 Mean   : 0.0000   Mean   : 0.000
```

```
   3rd Qu.: 0.2699    3rd Qu.: 0.246
   Max.   : 0.4723    Max.    : 0.456
> # compute empirical logit
> TargetFix$elog <- with(TargetFix,
                      log((sumFix+0.5) / (N-sumFix+0.5)))
> # compute weights
> TargetFix$wts <- with(TargetFix,
                      1/(sumFix+0.5) + 1/(N-sumFix+0.5))
> # fit full model
> TargetFix.elog <- lmer(elog ~ (ot1+ot2+ot3)*Condition +
                      (ot1+ot2+ot3 | Subject) +
                      (ot1+ot2+ot3 | Subject:Condition),
                   control=lmerControl(optimizer="bobyqa"),
                   data=TargetFix, weights=1/wts,
                   REML=FALSE)
> # get p-values for parameter estimates
> TargetFixElog.coefs <-
            data.frame(coef(summary(TargetFix.elog)))
> TargetFixElog.coefs$p <-
            2*(1-pnorm(abs(TargetFixElog.coefs$t.value)))
> TargetFixElog.coefs
                  Estimate Std..Error  t.value          p
(Intercept)      -0.113221   0.022365 -5.06248 4.1384e-07
ot1               2.725035   0.101688 26.79789 0.0000e+00
ot2              -0.544865   0.058374 -9.33410 0.0000e+00
ot3              -0.302512   0.037358 -8.09756 6.6613e-16
ConditionLow     -0.247764   0.030934 -8.00957 1.1102e-15
ot1:ConditionLow  0.032470   0.112802  0.28785 7.7346e-01
ot2:ConditionLow  0.688879   0.081721  8.42963 0.0000e+00
ot3:ConditionLow -0.084307   0.045440 -1.85533 6.3549e-02
> # plot data with full model fit
> p5 <- ggplot(TargetFix, aes(Time, elog, shape=Condition)) +
    stat_summary(fun.y=mean, geom="point") +
    stat_summary(fun.data=mean_se, geom="errorbar") +
    stat_summary(aes(y=fitted(TargetFix.elog),
                  linetype=Condition),
                fun.y=mean, geom="line") +
    theme_bw(base_size=10) +
    labs(x="Time since word onset (ms)",
        y="Fixation empirical logit",
        shape="Word\nFrequency", linetype="Word\nFrequency")
> # save plot as a PDF file
> ggsave("TargFixElogFit.pdf", p5,
        width=4.5, height=3, dpi=300)
```

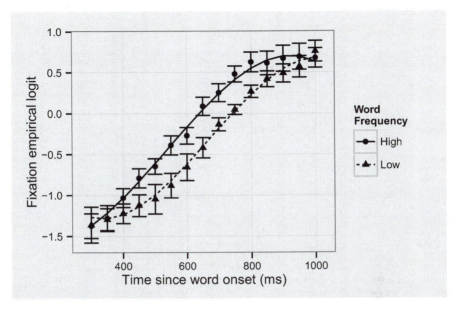

FIGURE 8.5

Observed target fixation empirical log-odds (symbols, error bars indicate ±SE) and growth curve model fits (lines) for effect of word frequency on the time course of spoken word recognition.

Example write-up:

> Growth curve analysis (Mirman, 2014) was used to analyze the time course of fixation from 300ms to 1000ms after word onset (i.e., from the earliest word-driven fixations to when target fixations had plateaued). The empirical logit transformation (Barr, 2008) was used to accommodate the categorical nature of the data (fixating the target picture or not) in a way that is robust to values at or near the boundaries (0 and 1). The overall time course of target fixations was captured with a third-order (cubic) orthogonal polynomial with fixed effects of condition (low vs. high frequency) on all time terms, and participant and participant-by-condition random effects on all time terms. The high frequency condition was treated as the reference (baseline) and relative parameters estimated for the low frequency condition. Statistical significance (*p*-values) for individual parameter estimates was assessed using the normal approximation (i.e., treating the *t*-value as a *z*-value). All analyses were carried out in R version 3.0.2 using the lme4 package (version 1.0-5).

> The data and model fits are shown in Figure 8.5. There was a significant effect of frequency on the intercept (*Estimate* = −0.248,

$SE = 0.0309$, $p = 1.11e - 15$), indicating overall higher fixation probability for the high frequency words than the low frequency words. There was also an effect of frequency on the quadratic term ($Estimate = 0.689$, $SE = 0.0817$, $p = 0$), reflecting faster recognition of high frequency words than low frequency words. Frequency did not have significant effects on the linear or cubic terms.

8.6   Individual differences as fixed effects

The data for this example come from a longitudinal study of English reading development in 181 children (from the ELDEL project, see Caravolas et al., 2012, 2013, and http://www.eldel.eu/). The example will show how to test the effect of four predictors (verbal span, letter knowledge, rapid naming, and phoneme awareness) that were measured at the start of the study on individual differences in the rate of learning to read.

```
> # Examine the data to check variable names and types
> summary(ELDEL)
         id           wdspan1            lk1                ran1
 ABBTUS :    6    Min.   :1.00    Min.   :-1.450    Min.   :-1.687
 ABIARM :    6    1st Qu.:2.00    1st Qu.:-0.114    1st Qu.:-0.623
 ABIHAR :    6    Median :3.00    Median : 0.343    Median :-0.115
 ABIJON :    6    Mean   :2.61    Mean   : 0.409    Mean   : 0.196
 AIDGRI :    6    3rd Qu.:3.00    3rd Qu.: 0.941    3rd Qu.: 0.802
 AISELE :    6    Max.   :4.00    Max.   : 2.102    Max.   : 3.684
 (Other):1050
       pa1               pwmcor           Month
 Min.   :-1.39468   Min.   : 1.0    Min.   : 0.0
 1st Qu.:-0.00416   1st Qu.: 8.0    1st Qu.: 4.0
 Median : 0.62316   Median :13.0    Median :13.0
 Mean   : 0.52247   Mean   :15.9    Mean   :13.7
 3rd Qu.: 1.04767   3rd Qu.:23.0    3rd Qu.:22.0
 Max.   : 1.88485   Max.   :61.0    Max.   :30.0
                    NA's   :82
> # Fit a model with an overall development effect (Month) and
> #    effects of all predictors on baseline reading
> #    performance (the intercept)
> ELDEL.intercepts <- lmer(pwmcor ~ Month + wdspan1 +
                              lk1 + ran1 + pa1 +
                              (Month | id),
                  data=ELDEL, REML=FALSE)
> # Evaluate effects of individual predictors on baseline
```

```
> #    performance
> ints <- drop1(ELDEL.intercepts, test="Chisq")
> # Fit full model
> ELDEL.full <- lmer(pwmcor ~ Month *
                                (wdspan1 + lk1 + ran1 + pa1) +
                                (Month | id),
                        data=ELDEL, REML=FALSE)
> # Evaluate effects of individual predictors on rate of
> #    reading development
> full <- drop1(ELDEL.full, test="Chisq")
> # Compute median split of letter knowledge for visualization
> ELDEL$LK <- factor(ELDEL$lk1 >= median(ELDEL$lk1),
                        levels=c("FALSE", "TRUE"),
                        labels=c("Low", "High"))
> # Plot effect of letter knowledge on reading development
> p6 <- ggplot(subset(ELDEL, !is.na(pwmcor)),
                aes(Month, pwmcor, shape=LK)) +
    stat_summary(fun.y=mean, geom="point") +
    stat_summary(fun.data=mean_se, geom="errorbar", width=1)+
    stat_summary(aes(y=fitted(ELDEL.full), linetype=LK),
                    fun.y=mean, geom="line") +
    theme_bw(base_size=10) +
    labs(y="Picture-Word Matching Score",
        shape="Letter\nKnowledge",
        linetype="Letter\nKnowledge") +
    theme(legend.position=c(0,1), legend.justification=c(0,1),
        legend.background=element_rect(color="black",
                                        fill="white"))
> # save plot as a PDF file
> ggsave("ELDEL.pdf", p6, width=4.5, height=4, dpi=300)
```

Example write-up:

The picture-word matching performance data were fit starting from a base linear growth model that modeled overall reading development with an intercept term (average reading ability at study start) and a linear slope term (rate of reading development over the 30 months of the study) and random effects of participants on the intercept and slope. Because the predictors of interest (verbal span, letter knowledge, rapid automatized naming, and phoneme awareness) were all moderately correlated with one another, a backwards elimination strategy was used to evaluate their unique contributions. To examine the effects of these critical predictors on initial reading ability, all four fixed effects were added to the base model, then each one was removed individually (leaving the other three in the model) and its unique effect evaluated by the reduction in

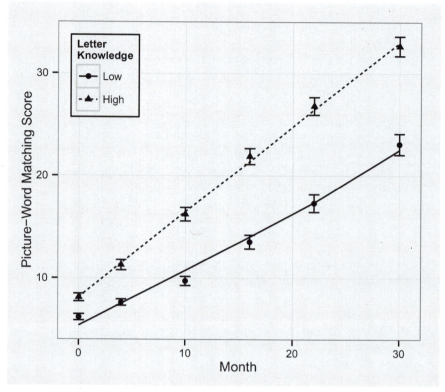

FIGURE 8.6
Observed development of reading ability (symbols, error bars indicate ±SE) grouped by median split on letter knowledge at study start. Lines indicate fit from full growth model.

model fit. Changes in model fit were evaluated using -2 times the change in log-likelihood, which is distributed as χ^2 with degrees of freedom equal to the number of parameters added (which was 1 for all comparisons). To examine the effects of the critical predictors on rate of reading development, a full model was constructed that contained effects of all of the predictors on the intercept and linear slope; then the effects of each of the four critical predictors on the slope was removed individually and the reduction in model fit evaluated. All analyses were carried out in R version 3.0.2 using the lme4 package (version 1.0-5).

The complete results of the model comparisons are shown in Table 8.2. Reading ability at study start was significantly predicted by letter knowledge, rapid automatized naming (RAN), and phoneme awareness, but not by verbal span. The rate of reading develop-

TABLE 8.2
Model Comparison Results Evaluating Effects of Removing Single Parameters on Model Fit

	Intercept: Chisq	p	Slope: Chisq	p
Verbal Span	0.153	0.696	0.513	0.474
Letter Knowledge	14.487	<0.001	13.241	<0.001
RAN	9.824	0.002	8.080	0.004
Phoneme Awareness	7.352	0.007	1.084	0.298

ment was significantly predicted by letter knowledge and rapid automatized naming, but not by verbal span or phoneme awareness. To visualize the effect of letter knowledge (which was the strongest predictor of both initial reading ability and rate of reading development), Figure 8.6 shows the observed data and model fit, with the participants median-split into high and low letter knowledge groups (all predictors were treated as continuous variables in the analyses; this discrete grouping was only used to create a simpler visual representation of the effects).

8.7 Individual differences as random effects

These data are from an eye-tracking experiment investigating the time course of activation of function and thematic knowledge in 17 participants with left hemisphere stroke (Kalénine, Mirman, & Buxbaum, 2012). Participants were presented with four pictures of familiar objects, heard a spoken word, and had to click on the matching object. One of the distractor objects was related to the target either because it serves a similar function (e.g., *broom - sponge*) or because it is typically used together with the target (e.g., *broom - dustpan*). The critical measure was fixation proportions on the related competitor relative to unrelated distractors in successive 50ms time bins from 500ms to 2000ms after word onset.

```
> # inspect the data
> summary(FunctThemePts)
      subj          Condition          Object           Time
 206    : 486   Function:4113   Target    :2733   Min.   :-1000
 281    : 486   Thematic:4086   Competitor:2733   1st Qu.:    0
 419    : 486                   Unrelated :2733   Median :  1000
 1088   : 486                                     Mean   :  1000
 1238   : 486                                     3rd Qu.:  2000
 1392   : 486                                     Max.   :  3000
```

```
(Other):5283
     timeBin          meanFix              sumFix                N
Min.   : 0    Min.    :0.0000    Min.    : 0.00    Min.    :12.0
1st Qu.:20    1st Qu.:0.0312    1st Qu.: 1.00    1st Qu.:15.0
Median :40    Median :0.1250    Median : 2.00    Median :16.0
Mean   :40    Mean    :0.1777    Mean    : 3.26    Mean    :15.4
3rd Qu.:60    3rd Qu.:0.2500    3rd Qu.: 5.00    3rd Qu.:16.0
Max.   :80    Max.    :1.0000    Max.    :16.00    Max.    :16.0
```

```
> # subset data for analysis
> FunctThemePts.gca <- subset(FunctThemePts,
                              Time >= 500 & Time <= 2000 &
                              Object != "Target")
> # adjust timeBin variable to start at 1
> FunctThemePts.gca$timeBin <- FunctThemePts.gca$timeBin - 29
> # make 4th-order orthogonal polynomial
> t <- poly((unique(FunctThemePts.gca$timeBin)), 4)
> # append it to data frame
> FunctThemePts.gca[, paste("ot", 1:4, sep="")] <-
                          t[FunctThemePts.gca$timeBin, 1:4]
> # fit separate models for Function and Thematic competition
> m.funct <- lmer(meanFix ~ (ot1+ot2+ot3+ot4)*Object +
                            (ot1+ot2+ot3+ot4 | subj) +
                            (ot1+ot2 | subj:Object),
                data=subset(FunctThemePts.gca,
                            Condition=="Function"),
                control=lmerControl(optimizer="bobyqa"),
                REML=FALSE)
> m.theme <- lmer(meanFix ~ (ot1+ot2+ot3+ot4)*Object +
                            (ot1+ot2+ot3+ot4 | subj) +
                            (ot1+ot2 | subj:Object),
                data=subset(FunctThemePts.gca,
                            Condition=="Thematic"),
                control=lmerControl(optimizer="bobyqa"),
                REML=FALSE)
> # extract subject-by-object random effects
> # Function condition
> re.id <- colsplit(row.names(ranef(m.funct)$"subj:Object"),
                    ":", c("Subject", "Object"))
> re.funct <- data.frame(re.id, ranef(m.funct)$"subj:Object")
> # Thematic condition
> re.theme <- data.frame(
            colsplit(row.names(ranef(m.theme)$"subj:Object"),
                    ":", c("Subject", "Object")),
            ranef(m.theme)$"subj:Object")
> # compute effect sizes
```

```
> ES.funct <- ddply(re.funct, .(Subject), summarize,
    Function_Intercept = X.Intercept.[Object=="Competitor"] -
                         X.Intercept.[Object=="Unrelated"],
    Function_Linear = ot1[Object=="Competitor"] -
                      ot1[Object=="Unrelated"])
> ES.theme <- ddply(re.theme, .(Subject), summarize,
    Thematic_Intercept = X.Intercept.[Object=="Competitor"] -
                         X.Intercept.[Object=="Unrelated"],
    Thematic_Linear = ot1[Object=="Competitor"] -
                      ot1[Object=="Unrelated"])
> # combine effect size estimates
> ES <- merge(ES.funct, ES.theme)
> # test the correlations
> # intercept term
> cor.test(ES$Function_Intercept, ES$Thematic_Intercept)
        Pearson's product-moment correlation

data:  ES$Function_Intercept and ES$Thematic_Intercept
t = -2.3602, df = 15, p-value = 0.03223
alternative hypothesis: true correlation is not equal to 0
95 percent confidence interval:
 -0.80075 -0.05300
sample estimates:
      cor
-0.52039
> # linear term
> cor.test(ES$Function_Linear, ES$Thematic_Linear)
        Pearson's product-moment correlation

data:  ES$Function_Linear and ES$Thematic_Linear
t = -3.3571, df = 15, p-value = 0.004322
alternative hypothesis: true correlation is not equal to 0
95 percent confidence interval:
 -0.86372 -0.25445
sample estimates:
      cor
-0.65499
```

Example write-up:

The time course of distractor fixations from 500ms to 2000ms after word onset was modeled using growth curve analysis (Mirman, 2014) with fourth-order orthogonal polynomials. Separate models were fit for the function and thematic conditions with fixed effects of object (related competitor vs. unrelated distractor) on

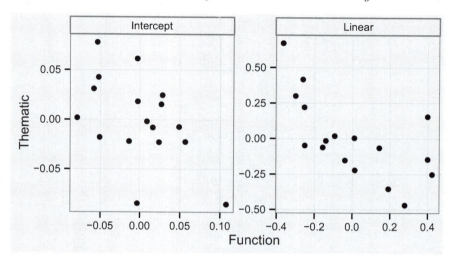

FIGURE 8.7
Scatterplots of individual function and thematic competition effect sizes on
intercept (left) and linear (right) terms.

all time terms, participant random effects on all time terms, and
participant-by-object random effects on the intercept, linear, and
quadratic time terms. The participant-by-object random effect estimates were used to compute individual participant effect sizes
estimates by, for each participant, subtracting the random effect
estimate for the unrelated distractor from the estimate for the related competitor. The intercept term captures overall differences in
fixation proportions between the related and unrelated distractors
and the linear term captures differences in the slope of the decrease
in distractor fixations, so these terms were most directly related to
competition effect sizes (the higher-order terms capture differences
in curvature that were less directly relevant to competition effect
size in this study). Therefore the effect sizes were estimated using
the intercept and linear term random effects only, from the separate function and thematic condition models. All analyses were
carried out in R version 3.0.2 using the lme4 package (version 1.0-
5).

Figure 8.7 shows scatterplots of individual participant function
and thematic competition effect sizes for the intercept (left panel)
and linear (right panel) terms. For the intercept term, there was
a significant negative correlation between function and thematic
condition effect sizes across the 17 participants ($r = -0.52$, $p =
0.032$), indicating that participants who showed larger function
competition effects tended to show smaller thematic competition

effects, and vice versa. There was a similar negative correlation pattern for the linear term ($r = -0.65$, $p = 0.0043$).

References

Baayen, H., Davidson, D., & Bates, D. (2008). Mixed-effects modeling with crossed random effects for subjects and items. *Journal of Memory and Language, 59*(4), 390–412. doi: 10.1016/j.jml.2007.12.005

Baguley, T. (2012). Calculating and graphing within-subject confidence intervals for ANOVA. *Behavior Research Methods, 44*(1), 158–175. doi: 10.3758/s13428-011-0123-7

Barr, D. J. (2008). Analyzing 'visual world' eyetracking data using multilevel logistic regression. *Journal of Memory and Language, 59*(4), 457–474. doi: 10.1016/j.jml.2007.09.002

Barr, D. J. (2013). Random effects structure for testing interactions in linear mixed-effects models. *Frontiers in Psychology, 4*(328), 1–2.

Barr, D. J., Gann, T. M., & Pierce, R. S. (2011). Anticipatory baseline effects and information integration in visual world studies. *Acta Psychologica, 137*(2), 201–207. doi: 10.1016/j.actpsy.2010.09.011

Barr, D. J., Levy, R., Scheepers, C., & Tily, H. J. (2013). Random effects structure for confirmatory hypothesis testing: Keep it maximal. *Journal of Memory and Language, 68*(3), 255–278. doi: 10.1016/j.jml.2012.11.001

Bates, E., Wilson, S. M., Saygin, A. P., Dick, F., Sereno, M. I., Knight, R. T., & Dronkers, N. F. (2003). Voxel-based lesion-symptom mapping. *Nature Neuroscience, 6*(5), 448–450. doi: 10.1038/nn1050

Bolker, B. (2013). *Algal (non)-linear mixed model example.* Retrieved Sept., 10, 2013, from http://rpubs.com/bbolker/3423

Brown, S., & Heathcote, A. (2003). Averaging learning curves across and within participants. *Behavior Research Methods, Instruments, & Computers, 35*(1), 11–21.

Caravolas, M., Lervåg, A., Defior, S., Seidlová Málková, G., & Hulme, C. (2013). Different patterns, but equivalent predictors, of growth in reading in consistent and inconsistent orthographies. *Psychological Science.* doi: 10.1177/0956797612473122

Caravolas, M., Lervåg, A., Mousikou, P., Efrim, C., Litavsky, M., Onochie-Quintanilla, E., . . . Hulme, C. (2012). Common patterns of prediction of literacy development in different alphabetic orthographies. *Psychological Science, 23*(6), 678–86. doi: 10.1177/0956797611434536

Cudeck, R., & Harring, J. R. (2007). Analysis of nonlinear patterns of change with random coefficient models. *Annual Review of Psychology, 58*, 615–637. doi: 10.1146/annurev.psych.58.110405.085520

Efron, B., & Morris, C. (1977). Stein's paradox in statistics. *Scientific Amer-*

ican, *236*(5), 119–127.

Francis, G. (2012). The psychology of replication and replication in psychology. *Perspectives on Psychological Science*, *7*(6), 585–594. doi: 10.1177/1745691612459520

Gelman, A., & Hill, J. (2007). *Data Analysis Using Regression and Multilevel/Hierarchical Models.* New York, NY, USA: Cambridge University Press.

Giacino, J. T., Whyte, J., Bagiella, E., Kalmar, K., Childs, N., Khademi, A., ... Sherer, M. (2012). Placebo-controlled trial of amantadine for severe traumatic brain injury. *The New England Journal of Medicine*, *366*(9), 819–26. doi: 10.1056/NEJMoa1102609

Grimm, K. J., Ram, N., & Hamagami, F. (2011). Nonlinear growth curves in developmental research. *Child Development*, *82*(5), 1357–1371. doi: 10.1111/j.1467-8624.2011.01630.x

Jaeger, T. F. (2008). Categorical data analysis: Away from ANOVAs (transformation or not) and towards logit mixed models. *Journal of Memory and Language*, *59*(4), 434–446. doi: 10.1016/j.jml.2007.11.007

Kalénine, S., Mirman, D., & Buxbaum, L. J. (2012). A combination of thematic and similarity-based semantic processes confers resistance to deficit following left hemisphere stroke. *Frontiers in Human Neuroscience*, *6*(106), 1–12.

Kalénine, S., Mirman, D., Middleton, E. L., & Buxbaum, L. J. (2012). Temporal dynamics of activation of thematic and functional action knowledge during auditory comprehension of artifact words. *Journal of Experimental Psychology: Learning, Memory, and Cognition*, *38*(5), 1274–1295. doi: 10.1037/a0027626

Kliegl, R. (2013). *Modeling time-accuracy functions with nlmer().* Retrieved Sept., 10, 2013, from http://read.psych.uni-potsdam.de/pmr2/index.php?view=article&id=89

Kliegl, R., Wei, P., Dambacher, M., Yan, M., & Zhou, X. (2011). Experimental effects and individual differences in linear mixed models: Estimating the relation of spatial, object, and attraction effects in visual attention. *Frontiers in Psychology*, *1*(238), 1–12. doi: 10.3389/fpsyg.2010.00238

Kriegeskorte, N., Simmons, W. K., Bellgowan, P. S. F., & Baker, C. I. (2009). Circular analysis in systems neuroscience: The dangers of double dipping. *Nature Neuroscience*, *12*(5), 535–540. doi: 10.1038/nn.2303

Magnuson, J. S., Dixon, J. A., Tanenhaus, M. K., & Aslin, R. N. (2007). The dynamics of lexical competition during spoken word recognition. *Cognitive Science*, *31*, 1–24.

Magnuson, J. S., Mirman, D., & Harris, H. D. (2012). Computational models of spoken word recognition. In M. J. Spivey, M. F. Joanisse, & K. McRae (Eds.), *The Cambridge Handbook of Psycholinguistics* (pp. 76–103). New York, USA: Cambridge University Press.

McClelland, J. L. (2009). The place of modeling in cognitive science. *Topics in Cognitive Science*, *1*(1), 11–38. doi: 10.1111/j.1756-8765.2008.01003.x

Mirman, D. (2014). *Growth Curve Analysis and Visualization Using R.* Florida, USA: Chapman & Hall/CRC.

Mirman, D., Dixon, J. A., & Magnuson, J. S. (2008). Statistical and computational models of the visual world paradigm: Growth curves and individual differences. *Journal of Memory and Language, 59*(4), 475–494. doi: 10.1016/j.jml.2007.11.006

Mirman, D., Holt, L. L., & McClelland, J. L. (2004). Categorization and discrimination of nonspeech sounds: Differences between steady-state and rapidly-changing acoustic cues. *Journal of the Acoustical Society of America, 116*(2), 1198. doi: 10.1121/1.1766020

Mirman, D., & Magnuson, J. S. (2009). Dynamics of activation of semantically similar concepts during spoken word recognition. *Memory & Cognition, 37*(7), 1026–1039. doi: 10.3758/MC.37.7.1026

Mirman, D., Magnuson, J. S., Graf Estes, K., & Dixon, J. A. (2008). The link between statistical segmentation and word learning in adults. *Cognition, 108*(1), 271–280. doi: 10.1016/j.cognition.2008.02.003

Mirman, D., Strauss, T. J., Brecher, A. R., Walker, G. M., Sobel, P., Dell, G. S., & Schwartz, M. F. (2010). A large, searchable, web-based database of aphasic performance on picture naming and other tests of cognitive function. *Cognitive Neuropsychology, 27*(6), 495–504. doi: 10.1080/02643294.2011.574112

Mirman, D., Yee, E., Blumstein, S. E., & Magnuson, J. S. (2011). Theories of spoken word recognition deficits in aphasia: Evidence from eye-tracking and computational modeling. *Brain and Language, 117*(2), 53–68. doi: 10.1016/j.bandl.2011.01.004

Oberauer, K., & Kliegl, R. (2006). A formal model of capacity limits in working memory. *Journal of Memory and Language, 55*(4), 601–626. doi: 10.1016/j.jml.2006.08.009

Pinheiro, J. C., & Bates, D. M. (2000). *Mixed-Effects Models in S and S-Plus.* New York: Springer.

Rafaeli, E., & Revelle, W. (2006). A premature consensus: Are happiness and sadness truly opposite affects? *Motivation and Emotion, 30*(1), 1–12. doi: 10.1007/s11031-006-9004-2

Scheepers, C., Keller, F., & Lapata, M. (2008). Evidence for serial coercion: A time course analysis using the visual-world paradigm. *Cognitive Psychology, 56*(1), 1–29.

Schwartz, M. F., & Brecher, A. R. (2000). A model-driven analysis of severity, response characteristics, and partial recovery in aphasics' picture naming. *Brain and Language, 73*(1), 62–91.

Schwartz, M. F., Kimberg, D. Y., Walker, G. M., Brecher, A. R., Faseyitan, O., Dell, G. S., ... Coslett, H. B. (2011). A neuroanatomical dissociation for taxonomic and thematic knowledge in the human brain. *Proceedings of the National Academy of Sciences, 108*(20), 8520–8524. doi: DOI:10.1073/pnas.1014935108

Singer, J. D., & Willett, J. B. (2003). *Applied Longitudinal Analysis: Modeling*

Change and Event Occurrence. New York: Oxford University Press.

Treisman, A. M., & Gelade, G. (1980). A feature-integration theory of attention. *Cognitive Psychology, 12*, 97–136.

Wagenmakers, E.-J., & Farrell, S. (2004). AIC model selection using Akaike weights. *Psychonomic Bulletin & Review, 11*(1), 192–196.

Index

aesthetics, 9, 10
    aes function, 9
    color, 12
    linetype, 10
    shape, 10
affect data, 15
aggregating data, 18
Akaike Information Criterion, 25
amantadine, 25
autocorrelated residuals, 39

Bayesian Information Criterion, 25
binary outcome variables, 104
binomial distribution, 104

contrast matrix, 97
contrasts, 87
    changing reference level, 92
    deviation coding, 87, 94
    dummy coding, *see* treatment
        coding
    sum coding, 87, 92
    treatment coding, 87, 89

`ddply` function, 81
dynamic consistency, 39

environment, 110
experimenter bias, 5, 8

facets, 12
    `facet_grid`, 12
    `facet_wrap`, 12
fixed effects, 23, 24, 27
formatting data, 15
    continuous to categorical, 126
    id variables, 17
    long, 17
    measure variables, 17

    `melt` function, 17
    wide, 15
functional form, 37
    considerations, 38

GCA, *see* growth curve analysis
geom, 9
    errorbar, 12, 34
    line, 10
    point, 10
    pointrange, 12
    ribbon, 66
`ggplot`, 9, 10
    `ggplot2` package, 9
    Grammar of Graphics, 9
    scales, 34
growth curve analysis, 3, 22
    linear example, 25
    mathematical overview, 22
    quadratic example, 51
    reporting results, 57, 143, 146,
        150, 152, 155, 157, 161

individual differences, 2, 8, 117
    as fixed effects, 118, 156
    as random effects, 127, 159
    individual models, 135
    plotting model fit, 125
    subsequent analyses, 137
inflection point, 46

likelihood ratio test, 25
linear
    change, 141
    growth model, 143
    slope, 143
`lmer`, 26
    simplified syntax, 30